Kelvin Probe for Surface Engineering: Fundamentals and Design

Kelvin Probe for Surface Engineering: Fundamentals and Design

A. Subrahmanyam

C. Suresh Kumar

CRC is an imprint of the Taylor & Francis Group,
an informa business

Ane Books Pvt. Ltd.

Kelvin Probe for Surface Engineering : Fundamentals and Design

A. Subrahmanyam and C. Suresh Kumar

© Authors

First Published in 2009 by

Ane Books Pvt. Ltd.
4821 Parwana Bhawan, 1st Floor
24 Ansari Road, Darya Ganj, New Delhi -110 002, India
Tel: +91 (011) 2327 6843-44, 2324 6385
Fax: +91 (011) 2327 6863
e-mail: anebooks@vsnl.net
Website: www.anebooks.com

For

CRC Press
Taylor & Francis Group
6000 Broken Sound Parkway, NW, Suite 300
Boca Raton, FL 33487 U.S.A.
Tel : 561 998 2541
Fax : 561 997 7249 or 561 998 2559
Web : www.taylorandfrancis.com

For distribution in rest of the world other than the Indian sub-continent

ISBN-10 : 1 42008 077 6
ISBN-13 : 978 1 42008 077 3

All rights reserved. No part of this publication may be reproduced, stored in a retrieval system, or transmitted in any form or by means, electronic, mechanical, photocopying, recording and/or otherwise, without the prior written permission of the publishers. This book may not be lent, resold, hired out or otherwise disposed of by way of trade in any form, binding or cover other than that in which it is published, without the prior consent of the publishers.

British Library Cataloguing in Publication Data
A catalogue record for this book is available from the British Library

Printed at Thomson Press, Delhi

Dedicated to
My Parents

Sri A. Bhima Sastry and Mrs Bhanumati
who laid the spirit of scientific enquiry in me.

A. Subrahmanyam

Preface

Surface engineering is the most powerful tool in the innovation of several highly functional surfaces spanning microelectronics, sensor technology, bio-medical engineering and many other emerging and strategic fields. The surfaces of metals and semiconductors look deceptively simple. Even after the development of highly sophisticated surface analytical equipment and dedicated investigations for five decades on semiconductor surfaces and interfaces, still there exist enough challenges. The Kelvin probe is the most powerful non-destructive technique to understand the active, passive and dynamic surfaces and interfaces / junctions. The Kelvin probe measures the surface work-function of semiconductor and metal surfaces. The associated techniques with the Kelvin probe are: Photo emission yield spectroscopy (PEYS) and Surface photovoltage spectroscopy (SPS), these techniques give wealth of information in the surface and interface physics / chemistry. The basics of Kelvin probe, though, are known for more than a hundred years, the instrumentation, somehow, has not caught the attention of many investigators. The recent advances in the signal recovery and the easy availability of ultra high vacuum, made the Kelvin probe instrumentation more accessible and now it is available commercially.

The Kelvin probe has been designed and built in the authors' laboratory with a focus to investgate semiconductor surfaces and interfaces / junctions. The main aim of writing the book is to share the experience for the benefit of the fellow scientists.

To the extent known to the authors, there is no text or monograph available yet on the fundamentals and design aspects of the Kelvin Probe equipment.

This book is a modest attempt :

 (i) to compile the basics of the Kelvin probe (Chapter 1) and to give clarity to the definition of the chemical potential and work function from the statistical thermodynamics point of view (Chapter 2)
 (ii) to elaborate various aspects of the design and several possibilities of vibrating reed mechanisms and to describe the design and fabrication details (Chapter 3)
 (iii) to explore the associated experimental techniques related to the Kelvin probe : photoemission yield spectroscopy (Chapter 4) and Surface Photovoltage spectroscopy (Chapter 5).

The contents covered in this book are not exhaustive but sufficient to understand and build the Kelvin probe equipment. It should be mentioned that the approach detailed in this book to fabricate the Kelvin probe is not unique, but it is fully functional and cost effective. The material presented in this book is at the elementary level; a basic knowledge of semiconductor surface physics and preliminary knowledge in electronic circuits is desired, but not essential. Several examples have been presented (most of these data are from the authors' laboratory) so that it becomes easier to follow and gain confidence. The authors are aware that a smooth reading of the text is hindered by the references, however, it is for the advantage of having more details, particularly in the design aspects.

It is the sincere hope that the book serves the community of scientists and technologists in the area of surface engineering.

The authors sincerely thank the Department of Science and Technology (DST), Government of India for providing the funds and necessary encouragement for the construction of the Kelvin probe. The support provided by the Indian Institute of Technology Madras, Chennai to carry out the work is gratefully acknowledged. The authors thank Dr Paul Benny who has not only initiated the first designs but also infused the level of confidence that "we can build the equipment" spirit. The authors acknowledge the contribution of all the people who have contributed in several ways in the design and building of the Kelvin probe equipment. The support and the understanding extended by Mrs Vanaja Subrahmanyam during the course of this book writing should be acknowledged as sincerely as a husband (AS) can. Authors thank sincerely the perseverance and patience shown by Mr Sunil Saxena of M/s Ane Books Pvt. Ltd. without whose impetus and encouragement, the book might not have seen the light of the day.

Every possible care has been taken to present the material as accurately as possible, but to err is human, there might be some mistakes.

A. Subrahmanyam
C. Suresh Kumar

Biography of Lord Kelvin

William Thomson is a child prodigy and is a rare breed of scientists. He was born on 26th June, 1824 in Belfast, Ireland. His mother died when he was six, and he moved with the family to Glasgow. James Thomson, his father, was a Professor of mathematics at the University of Glasgow and was his first teacher.

Thomson entered the University of Glasgow in 1841. Thomson went to Cambridge University in 1845 and did postgraduate work in Paris with Professor Regnault. Throughout his education, Thomson demonstrated excellence and published scholarly papers on mathematics, the first when he was 16. He became a Member of the Royal Society while he was 20!

He became Professor of Natural Philosophy at Glasgow University in 1846 and he held the position for almost fifty years. His work was mainly on the subject of heat and thermodynamics. His definition of absolute temperature scale in 1847 bears his name in the history of Physics. In 1851 he published ideas leading to the second law of thermodynamics. He changed the view of heat as being a fluid to an understanding of the energy of motion of molecules. The term "kinetic energy" was introduced by Thomson in 1856. Thomson worked with Joule to discover the Joule-Thomson effect, the foundation for the work on low temperature physics.

Thomson had published the paper: 'The Uniform Motion of Heat in Homogeneous Solid Bodies, and its Connection with the Mathematical Theory of Electricity,' (*Cambridge Mathematical Journal*, vol. iii., 1842). This analogy and understanding brought him success in laying the trans-Atlantic telephone line. He improved the design of the cables, even traveled on the ships supervising the laying of them. He exercised enormous perseverance with the transatlantic cable project despite problems, setbacks and the need to restart more than once. With the breadth of experience acquired, he became a famous consultant for the submarine cable projects.

His other major contribution is in the study of surface properties of metals. He is probably the first to visualize the non-destructive method of evaluating the metal surfaces with an electric probe: the Kelvin probe (the subject of this book). This experience prompted Kelvin's invention of the mirror galvanometer (patented, 1858) as a long distance telegraph receiver which could detect extremely feeble signals. He studied the electrical losses in cables, and improved the mariner's work with the invention of an improved gyro-compass, new sounding equipment, and a tide prediction with chart-recording machine. He also introduced Bell's telephone into Britain.

Thomson published more than 600 scientific papers and filed a total of 70 patents. He was the president of the Royal Society from 1890 to 1895.

In 1866 he was Knighted for his achievements in the submarine cable laying. In 1892 he was raised to the peerage as Baron. He chose the title: Kelvin of Largs, from the Kelvin River, near Glasgow. He was Britain's first scientific peer.

He died on 17th December, 1907. He was buried next to Isaac Newton in Westminster Abbey.

"When you can measure what you are speaking about and express it in numbers, you know something about it."

Lecture to the Institution of Civil Engineers, 3rd May, 1883

Lord Kelvin

Contents

Preface	*vii*
Biography of Lord Kelvin	*ix*
Acronyms/Abbreviation	*xv*
Symbols Used	*xvii*

Chapter 1: Introduction to Surfaces and Interfaces   1-20

1.1	Introduction	1
1.2	Basics of Surfaces and Surface Science	3
1.3	The Structure of Surfaces	4
1.4	The Ionization Energy and Heat of Formation in Surfaces	6
	1.4.1 Ionization Energy	6
	1.4.2 Heat of Formation	6
1.5	Mobility of Atoms on the Surface	6
1.6	Surface Band Structure and Surface States	7
1.7	Interaction of Adsorbed Species and Adsorbate-Induced Changes in Surface Electronic Properties	9
	1.7.1 Monovalent Adatom	10
1.8	Adsorbate-induced Work Function Changes	11
1.9	An Outline of Surface Characterization Techniques	13
1.10	Experimental Techniques for Surface Structure Evaluation	14
1.11	Chemical Composition and Bonding	14
1.12	Electronic and Optical Properties	15
1.13	Kelvin Probe Applied to Surface Studies	18
1.14	A Brief Record and Versatility of Kelvin Probe	18

Chapter 2: The Thermodynamics of Surfaces and   21-44
Definition of Work Function and Measurement

2.1.	Introduction to Thermodynamics of Charges on the Surfaces	21
2.2.	Chemcial Potential and Contact Potential Difference	22

	2.2.1 Surface Energy	22
	2.2.2 Chemical Potential	24
2.3	Chemical Potential of Few-electron System or Atomic Chemical Potential	26
2.4	Fermi Energy and Chemical Potential in Semiconductors	27
2.5	Electrochemical Potential	29
2.6	Electrochemical Potential Distribution with Electrical Current and Temperature Distribution	30
2.7	Definition of Work Function	30
	2.7.1 Work Function Based on Electro-chemical Potential	30
	2.7.2 Work Function Based on Contact Potential Difference	32
	2.7.3 Work Function Based on Thermionic / Field Emission	32
	2.7.4 Note on the Diversity of Work Function	32
2.8	Work Function: Band Model	33
2.9	The Vacuum Level	33
2.10	Negative Work Function	34
2.11	Surface and Bulk Work Function of Metals and Semiconductors	35
	2.11.1 Work Function Measurements Based upon Diode Method	35
	2.11.2 Field Emission Technique of Measuring Work Function	36
	2.11.3 Photoelectric Method of Measuring Work Function	37
2.12	Contact potential Difference : Metal-Metal System	38
2.13	Principle of Measuring CPD	40
	2.13.1 Metal–semiconductor System	41
2.14	Non-uniformity of Real Surface and Averaging of the Kelvin Probe	42
2.15	Anisotropy of the Work Function	42

Chapter 3: Design of the Kelvin Probe and the Methods of Measurement of Contact Potential Difference (CPD)   45-82

3.1	Introduction	45
3.2	The Design Aspects of the Kelvin Probe: The Vibrating Capacitor	47
	3.2.1 Ideal Parallel Plate Capacitor Geometry	47
	3.2.2 Effect of Fringe Field and Non-parallelism	49
	3.2.3 Stray Capacitance Effect	51
	3.2.4 Non-uniformity of Work Function	51
3.3	The Modulation Methods and Noise Reduction Techniques: An Outline of Literature	51
3.4	Kelvin Probe Design	52
	3.4.1 Kelvin Probe Head Mechanical Design	53
	3.4.2 Vibration Characteristics of the Probe	53

3.5	Description of the Measuring Circuit	57
	3.5.1 Conversion of the Displacement Current of Vibrating Capacitor into Voltage	57
	3.5.2 The Preamplifier	59
	3.5.3 Analysis of the Preamplifier Output	60
	3.5.4 Harmonic Content of the Output Signal	62
3.6	Method of CPD Measurement	67
	3.6.1 Off-null Method	67
	3.6.2 Feed Back Loop System	68
	3.6.3 Effect of Noise	70
3.7	Spacing Dependence of CPD	71
3.8	The Reference Electrode	73
3.9	Vacuum System	74
3.10	Description of the Sample Holder	75
3.11	Scanning Kelvin Probe	75
	3.11.1 Electrochromic Properties of Tungsten Oxide	78
	3.11.2 Electrical Potential Gradient in the Presence of Local Electric Current	80

Chapter 4: Photo-Emission Yield Spectroscopy (PEYS) 83-110

4.1	Introduction	83
4.2	Basics of Photoemission from Solids	84
	4.2.1 Photoelectric Work Function	84
	4.2.2 Photoemission Process	85
	4.2.3 Experimental Techniques Based on Photoemission	86
4.3	Fowler's Theory of Threshold Photoemission from Metals	86
4.4	Band Structure Approach to Ionization Energy of Semiconductor Surfaces	87
4.5	Measurement of Photoelectric Threshold	89
	4.5.1 Experimental Set up	89
	4.5.2 Measurement Procedure	91
	4.5.3 Types of Samples	92
	4.5.4 Preparation of Reference Electrode in PEYS Experiments	93
4.6	Studies on Metallic Surfaces	93
	4.6.1 Gold Thin Film	94
	4.6.2 Gold Foil	97
	4.6.3 Silver Thin Film	99
	4.6.4 Molybdenum Thin Film	101
	4.6.5 Reference Electrode Work Function : Graphite	104
	4.6.6 Studies on GaAs (100) Surfaces	106

Chapter 5: Surface Photovoltage (SPV) Spetroscopy of Semiconductor Surfaces — 111-142

5.1	Introduction	111
5.2	A Brief Literature on SPV Technique	112
5.3	Basics of Surface Photovoltage Spectroscopy	113
5.4	An Outline of the Relations between the Excess Charge (Δp) and SPV	114
	5.4.1 Super-band Gap SPV	116
	5.4.2 Sub-band Gap SPV	116
	5.4.3 Principle of SPV Measurement Using Kelvin Probe	118
5.5	Surface Photovoltage Experimental Setup	119
	5.5.1 Samples for SPV Measurement	120
5.6	Results	121
	5.6.1 Super-band Gap SPV in GaAs, CdS and CdTe/Silicon Heterojunction	121
	5.6.2 Moderately Doped n-type GaAs (100) Surfaces	122
	5.6.3 Heavily Doped p-GaAs (100) Surfaces	130
	5.6.4 Semi-insulating GaAs (100) Surfaces	133
	5.6.5 GaAs p-n Junction	137

APPENDICES — 143-158

Appendix I	Surface States in Semiconductors	143
Appendix II	The Fourier Coefficients of $C_K(t)$	147
Appendix III	Specifications of Low Power FET –input Electrometer Grade Op Amp AD515AJ	153
Appendix IV	Theory of Large-Signal Surface Photovoltage	154
References		159-180
Index		181-182

Acronyms/Abbreviation

ADC	Analogue to Digital Convertor
AES	Auger Electron Spectroscopy
AFM	Atomic Force Microscopy
ARUPS	Anile Resolved Ultraviolet Photo Emission Spectroscopy
CNL	Charge Neutrality Level
CPD	Contact Potential Difference
DAC	Digital to Analog Convertor
DAS	Data Acquisition System
DFT	Density Functional Theory
DLTS	Deep Level Transient Spectroscopy
DOS	Density of States
DSO	Digital Storage Oscilloscope
EBIC	Electron Beam Induced Current
EFIRS	Electron Field Induced Raman Spectroscopy
HOPG	Highly Ordered Pyrolytic Graphite
HRELS	High Resolution Energy Loss Spectroscopy
ICL	Interface Control Layer
KPFM	Kelvin Probe Force Microscopy
KRIPES	K-resolved Inverse Photo Emission Spectroscopy
LDA	Linear Density Approximation
LEED	Low Energy Electron Diffraction
LIA	Lock in Amplifier
MBE	Molecular Beam Epitaxy
MIS	Metal Insulator Semiconductor
ML	Mono Layer
NEA	Negative Electron Affinity
NRA	Nuclear Reaction Analysis
PES	Photo Electron Spectroscopy

PEYS	Photo Emission Yield Spectroscopy
PL	Photo Luminescence
PMR	Polarization Modulated Reflectivity
PMT	Photomultiplier Tube
RAS	Reflection Anisotropy Spectroscopy
REED	Reflection High Energy Diffraction
RT	Room Temperature
SAM	Self Assembled Microscopy
SCL	Space Charge Layer
SDR	Surface Differential Reflectivity
SEXAFS	Surface Extended X-ray Absorption Fine Structure
SPS	Surface Photo Voltage Spectra
SPS	Surface Photo voltage Spectroscopy
SPV	Surface Photo Voltage
STM	Scanning Tunneling Microscope
SVM	Scanning Voltage Microscopy
TPD	Temperature Programmed Desorption
UHV	Ultra High Vaccum
UPS	Ultraviolet Photo Spectroscopy
UPS	Ultraviolet Photoemission Spectroscopy
VBM	Valence Band Maximum

Symbols Used

ϕ_B	Barrier Height
ΔH_m	Activation Energy of Migration
μ	Chemical Potential
A	Helmholtz Energy
C_i	Stray Capacitance
C_k	Kelvin Probe Mean Capacitance
D	Dipole Operator
d_{cov}	Covalent Bond Length
E_{kin}	Kinetic Energy of the Emitted Photo Electron
E_{VBM}	Energy of the Valence Band Maximum
E_{vac}	Vacuum Reference Level
E_{vs}	Valence Band at the Surface
G	Gibbs Energy
H	Enthalpy
H_F	Heat of Formation
H_R	Heat of Reaction
I	Ionization Energy
N_{ad}	Adsorbate Induced Surface Dipoles
P	Dipole Movement
Qsc	Intrinsic Surface State Density
Qss	Minimum Surface State Density
$S(\phi)$	Sticking Coefficient
S^*	Entropy Flow per Particle
U	Total Internal Energy
V_D	Dember Voltage
X_A	Electronegativity of Atom A
X_B	Electronegativity of Atom B
α_{ad}	Polarizability of the Adatom

γ	Surface Tension
μ^P	Chemical Potential of Positron
σ	Electrical Conductivity
Φ_-	Electron Work Function
Φ_+	Positron Work Function
Φ^p	Work Function at Photothreshold
χ	Electron Affinity

Chapter 1

Introduction to Surfaces and Interfaces

1.1 INTRODUCTION

Surface science and engineering is an essential and inevitable part of our everyday life and governs a majority of strategic and most advanced applications. The surface science addresses the issues related to the solid state electronics, corrosion, sensors, tribological coatings, heterogeneous catalysis, optical properties of surfaces, bio-medical devices, optical and magnetic non-volatile memories and functional properties and Self-Assembled Monolayers (SAMs), just to name a few. The studies in the fields of surface and interface science and surface engineering (modification of solid surfaces) are making tremendous strides in bringing technological breakthroughs with new and novel devices. The surface properties of solids are influenced to a large extent by the underlying bulk properties. With the emergence of low dimensional solids, ultra-thin films of nanometer grain sizes, organic semiconductors and the Self-Assembled Monolayers (SAMs), the surface studies are extremely important not only for the fundamental understanding of the surface kinetics and surface stability but also for the functionalization of the surface for most advanced applications in several areas. A detailed understanding of surfaces and interfaces enhances the ability to predict and control their properties and to design newer and better interfaces. Tailored surfaces and interfaces are always a dream for emerging technologies.

Sophisticated surface-sensitive experimental techniques like Low Energy Electron Diffraction (LEED), X-ray Photoelectron Spectroscopy (XPS), Ultraviolet Photoemission Spectroscopy (UPS), *etc.*, have been successfully employed to characterize the structural, chemical and electronic properties of semiconductor surfaces (Mönch, 1993; Vickerman, 1997). However, all these probes disturb the virginity of the surface.

The Kelvin probe is an equally competent, non-destructive, more powerful and less expensive technique to study the electronic properties of surfaces and interfaces. The Kelvin probe basically measures the difference in the work function between the sample surface under investigation and a reference surface/electrode. In this book, the basic principles and the design details of the Kelvin probe are explained. Though, most of the examples in this book are concerned with and limited to the semiconductor surfaces and interfaces, the Kelvin probe technique has much wider reach in addressing almost any active or passive surface.

The Kelvin probe was first proposed by Lord Kelvin in 1898 to investigate the contact potential difference (CPD) between metallic surfaces (Kelvin, 1898).

This chapter gives a brief introduction of the physics of the surfaces and the surface kinetics of adsorbates. The topics covered are: the basics of surface states in semiconductors, structure of surfaces, ionization energy and heat of formation in surfaces, mobility of adatoms on the surface, interaction of adsorbed species and adsorbate-induced changes in surface electronic properties and work function changes, Kelvin probe applied to surface studies and a brief chronology of the Kelvin probe technique.

According to crystallography, the ideal "surface" itself is a defect due to the unsaturated bonds. The "real" surface is much more complex in terms of structure, morphology and the reaction kinetics. For example, for a given material and a given surface orientation, a variety of different surface structures/morphologies can exist (Fig. 1.1). The surface structure not only depends on the preparation, treatment after preparation and conditions during investigation but also on the reactivity of the surface and the available surface energy for any reaction, chemical or physical. This leads to very interesting and complex structural, chemical, electrical, magnetic and optical properties. It is important to develop perspective in terms of real surfaces. The repeatability and reproducibility of the surface is altogether another aspect of challenge in the surface science and engineering.

The properties of surfaces in metals and semiconductors are very much different (and are less understood) from the properties of their bulk (Many et al., 1965; Mönch, 1993). The extensive and exhaustive investigations on the surface and interface properties of semiconductor materials have immensely contributed (and still contributing) to the rapid growth of micro and nano electronics and photonics. With increasing demand for semiconductor device miniaturization, a careful tailoring of their surface and interface properties became a necessity. From the device point of view, the most extensively studied materials are Silicon (Si) and Gallium Arsenide (GaAs). However, a lot is still to be done for the GaAs and many new semiconductor materials, mainly to understand and improve the junctions: metal-insulator-semiconductor (MIS), metal-semiconductor (MS) and heterostructures. These junctions are the building blocks for many devices: microwave monolithic integrated circuits (MMIC's) and optoelectronic integrated circuits, just to name a few (Wieder,1980). GaAs surface is a negative electron affinity (NEA) surface, characterized by a high probability of thermal electron emission into vacuum (Feuerbacher *et al.*, 1978).

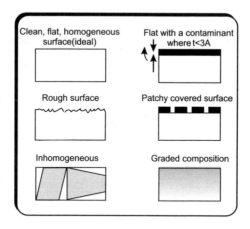

Fig. 1.1 Possible types of surfaces commonly encountered.

Unlike silicon, GaAs surface cannot be easily passivated either by oxidation or by chemical treatments. Nevertheless, certain passivation schemes involving interface control layer (ICL) leads to greater reduction of interface states in GaAs and hence the realization of device structures (Negoro *et al.*, 2003).

With the increasing demand for semiconductor device miniaturization, surface effects require careful tailoring of their surface properties. Similar are the challenges in the emerging areas of functionalization including organic semiconductor/metal interfaces, bio-functionalization of coatings, Self-assembled monolayers on a variety of substrates, the single and multi-walled nano tubes and for surface Plasmon assisted devices. For all the surface and interface studies, the fundamental requirement is a technique that extracts maximum information with minimum or no surface/interface damage.

1.2 BASICS OF SURFACES AND SURFACE SCIENCE

The surface of metals and semiconductors are deceptively simple (Many *et al.*, 1965; Mönch, 1993). There is a vast amount of literature available on the solid surfaces and surface characterization techniques (Reviere and Myhra, 1988; Czanderna 1989). The abrupt termination of the three dimensional (3D) ideal periodic crystal lattice and the periodic potential at the surface gives rise to a complex surface structure and associated localized electronic states (surface states). Apart from this "ideal defect", the real surfaces have much more complexities in terms of a variety of 'defects' (Fig. 1.2). The unsaturated atomic orbits or dangling bonds available at the surface are generally unstable and energetically less favourable. In order to form energetically more stable configuration (thermodynamic equilibrium), the surface undergoes simple relaxation or complete reconstruction of their surface atom position from their ideal truncated bulk position. Both phenomena: relaxation and reconstruction, can also take place.

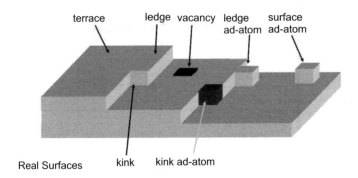

Fig. 1.2 The defects that can occur on real surfaces.

This leads to changes in the bond angle and relative positions of the atoms at the surface and a few monolayers below the surface. The other possibility for the high energy surface is to interact with the available external atoms which can stick on to the surface (called the 'adatoms') thus minimizing the surface energy. The 'adsorption kinetics' depend upon the nature of the surface and the available surface energy. Clean surfaces can be created by cleaving the bulk crystal in ultra high vacuum (UHV) or by removing the adsorbed layer in UHV.

1.3 THE STRUCTURE OF SURFACES

The dangling bonds available at the surface interact with various species in the surrounding giving rise to adsorbate surfaces. On both clean and adsorbate-covered surfaces, the saturation of dangling bonds are generally accompanied by local deformation of bond angles (which induce strain in the lattice) while the bond length remains almost unchanged (Brattain *et al.*, 1953). Because of the two dimensional translation symmetry of the surface, the corresponding surface electronic structure forms two-dimensional bonds.

The atomic structure can be calculated once the electronic structure is known since electrons and their states are essential for the forces determining the atomic positions. Fig. 1.3 shows the typical example of the atomic arrangement and Brillouin Zone of the bulk and (1×1) unit mesh of the bulk (100) plane of the zincblende structure.

The two structural problems: relaxation and reconstruction must be solved in a self-consistent manner. Generally, for semiconductor surfaces, bulk impurity concentration upto $10^{17}/cm^3$ has little effect on the geometric structures and on the surface states, but they do affect the surface barrier heights and related properties.

Surface structure of cleavage planes, and technologically important low index planes of both elemental and compound semiconductors and their subsequent changes during various surface treatments such as annealing, ion bombardment, interaction of foreign atoms (initial stages of Schottky barrier formation, oxidation, hetero-junction growth), wet chemical treatment, etc. have been investigated extensively from the early days of LEED to modern real space imaging techniques such as Scanning Tunneling Microscopy (STM) (Many *et al.*, 1965; Brillson, 1982; Mönch, 1993).

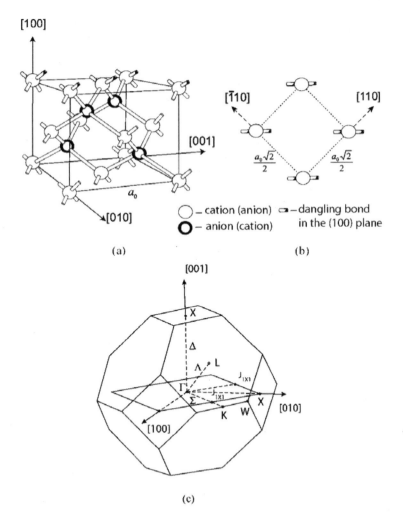

Fig. 1.3 (a) Unit cell of the zincblende structure, (b) Atomic arrangement in the (1×1) unit mesh of the bulk (100) plane, and (c) Bulk Brillouin Zone with the (001)-(1×1) surface Brillouin Zone shown as insert. Symbols represent high-symmetry points in the Brillouin Zone.

1.4 THE IONIZATION ENERGY AND HEAT OF FORMATION IN SURFACES

1.4.1 Ionization Energy

The ionization energy is a characteristic of an ideal virgin surface. For example, the ionization energy (I) for a semiconductor surface may be defined (following the band theory) as the energy difference between the vacuum reference level (E_{vac}) and the top of the valence band at the surface (E_{vs}).

$$I = E_{vac} - E_{vs} = E_{vac} - E_{cs} + E_g = \chi + E_g \qquad \ldots 1.1$$

E_{cs} is the conduction band edge and χ is the work function and E_g is the band gap. The surface stoichiometry and structure give the main contribution to the variation in the ionization energy (Ranke, 1983). The ionization energy can be calculated using tight binding approach, density functional theory and *ab-initio* pseudopotentials (Alves, *et al.*, 1991). A good agreement has been obtained between the theoretical and experimental results (van Laar, *et al.*, 1977) for (110) surfaces of gallium and indium containing III-V semiconductors.

The surfaces of metals and semiconductors do possess dipole layers due to crystallographic aniostropy. Additional dipole layers, present due to the adsorbed species/atoms, significantly modify the ionization energy of metals and semiconductors. External electric field or internal electric field due to space charge layer can modify the surface dipole layer and hence the ionization energy (Fischer and Viljoen, 1971).

1.4.2 Heat of Formation

The magnitude of heat of formation, H_F is a measure of the semiconductor stability against chemical reaction. Hence, the materials with smaller values of H_F seem to be less stable and tend to form new chemical bonds in the interface region (termed as heat of reaction: H_R). Low barrier heights (ϕ_B) appear for reactive interface, whereas in the case of unreactive interfaces, the barrier height seems to approach a certain value for $\Delta H_R \to \infty$. The transition in ϕ_B occurs near $\Delta H_R = 0$. Weak ionic semiconductors tend to form Bardeen-like barriers by Fermi-level pinning and strong ionic compounds tend to form Schottky-like barrier.

1.5 MOBILITY OF ATOMS ON THE SURFACE

The virgin surface is attacked by external atoms either from the ambient or atoms intentionally added on to the surface (for example, as contacts). The surface mobility of arriving atoms may be described by a hopping frequency, which can be expressed as $v \exp\left(-\dfrac{\Delta H_m}{k_B T}\right)$. The frequency factor v typically amounts to $10^{12}/s$ and activation energy of migration, ΔH_m has been estimated as 0.3 to 0.6 eV for Al on GaAs (110). An isolated Al atom is thus expected to make 10^4 jumps/s at room temperature (RT= 300 K) but only one jump in 10^5 days at 100 K.

1.6 SURFACE BAND STRUCTURE AND SURFACE STATES

In the early days of understanding of the surfaces, Tamm (Tamm, 1932) and Shockley (Shockley, 1939) demonstrated theoretically the occurrence of surface states within the fundamental band gap of a crystal. The invention of germanium point-contact transistor has enhanced the fundamental understanding of the surface properties of semiconductors (Bardeen and Brattain, 1948). The Field effect, originally suggested for the field-effect transistor, has been an extremely fruitful tool for the fundamental investigation of the surface states (Shockley and Pearson, 1948). However, for wide band gap semiconductors like GaAs, because of the low intrinsic carier density $(n_i = 10^7 / cm^3)$, it is very difficult to interpret field effect data.

Bardeen attributed the deviation of experimental barrier heights in metal-semiconductor contacts from the Schottky-Mott rule to the existence of interface states (Bardeen, 1947). The thermally oxidized silicon-silicon dioxide (Si-SiO_2) interface (99.9999% of the interface bonds are defect free) has laid the foundations for the understanding of interface states.

Surface-induced defects and the structural alterations like the surface reconstruction can bring out detectable changes in the surface electronic structure, of the order of few tenths of an eV to about 1 eV. For covalent semiconductors, the lattice disruption at the surface is large. The crystal bonding in ionic semiconductors shows less disruption of the lattice potential near the surface resulting in a lower density of surface states.

A number of experimental and theoretical techniques have been developed for studying the surface state band structures of semiconductors. The surface band structure calculations are based on the model systems in the form of a finite cluster size, a semi-infinite slab, or a repeated slab geometry (Srivastava, 2000). The band structures of surface states on semiconductor surfaces are now routinely calculated using well-developed techniques, like, Density Functional Theory (DFT), Linear Density Approximation (LDA) calculations using *ab initio* pseudopotentials and a slab geometry (Alves *et al.*, 1991), self-consistent pseudopotential method, the tight binding method, scattering-theoretical approach, and Green's function approach. The popular experimental techniques employed for surface band structure are: angle resolved ultraviolet photoemission spectroscopy (ARUPS), k-resolved inverse photoemission spectroscopy (KRIPES) and scanning tunneling microscopy (STM). Theoretical energy dispersion of occupied and unoccupied surface states showed (Fig.1.4) closer agreement with dispersion of occupied surface obtained using ARUPS and KRIPES (Larsen *et al.*, 1981; Salmon and Rhodin, 1983). In ARUPS, the energy dispersion of the surface bands is directly derived from the variation of peak position as a function of emission angle. Using STM, the occupied and unoccupied dangling bond surface states are obtained in real space with atomic resolution by changing the polarity of the bias voltage applied between the tip and the sample under investigation. Polarization-modulated reflectivity (PMR) and surface differential reflectivity (SDR) techniques are also used to study the surface states well outside the band gap.

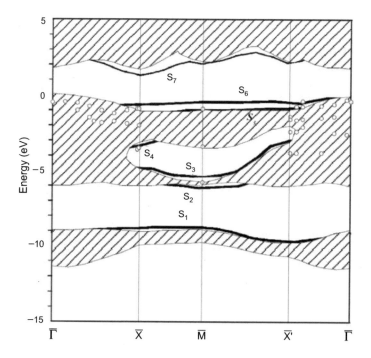

Fig. 1.4 Schematic of the electronic band structure of the clean InP (110) surface. Projected bulk spectrum is shown by hatched regions. Occupied and unoccupied surface states are shown by thick and thin curves, respectively. The angle-resolved photoemission results are indicated by open circles (Larsen *et al.*, 1981). The states, S_1 (s-like), S_4 (px-like), S_5 and S_6 (pz-like) are derived from the anion (P) atoms and the other states, S_2 (s-like), S_3 (s-like) and S_7 (spz-like), are derived from the cation (In) atoms. The orbital characters of these states are rather well known. The unoccupied surface energy level, S_7, is found to lie at an energy 1.5 eV above the bulk valence band edge at $\bar{X}$.

1.7 INTERACTION OF ADSORBED SPECIES AND ADSORBATE-INDUCED CHANGES IN SURFACE ELECTRONIC PROPERTIES

It is well-known that foreign atoms (adsorbates) on semiconductor surfaces induce surface dipoles in addition to the surface states. The defect creation and the charge transfer mechanism between the adsorbate atoms and the surface of the substrate may be evaluated by the work function measurements. The ionicities of covalent bonds between different atomic species are correlated with the electronegativities of the atoms involved (Pauling, 1932). Hannay and Smyth revised Pauling's original relation of ionocity as (Hannay and Smyth, 1946)

$$\Delta q_1 = 0.16|X_A - X_B| + 0.035|X_A - X_B|^2 \qquad ...1.2$$

where X_A and X_B are the atomic electronegativities of the atoms A and B forming the molecule.

In a simple point charge model, the atoms are charged by $+\Delta q_1 q$ and $-\Delta q_1 q$, where the more electronegative atom becomes negatively charged. So a diatomic molecule with $|X_A - X_B| \neq 0$ possesses dipole moment given by

$$p = \Delta q_1 q d_{cov} \qquad ...1.3$$

where d_{cov} is the covalent bond length. The above mentioned partial ionic character of diatomic molecule can be applied to adsorbates on semiconductor surfaces by using surface molecule model. The average electronegativity of the surface molecule is given by the geometric mean of the atomic values of their constituents. For binary semiconductors the average electronegativity can be taken as

$$X_{AB} = (X_A X_B)^{\frac{1}{2}} \qquad ...1.4$$

Based on the above model, adsorption electropositive elements like cesium ($X_{Cs} = 0.79$) on semiconductors reduces their ionization energy while electronegative elements like chlorine ($X_{Cl} = 3.2$) increases the ionization energy.

Adsorbate induced surface states will act as acceptors and donors when they are closer to the conduction band minimum and the valence band maximum, respectively. Depending on the wavefunction tail, the electronic charge may shift towards or away from the surface and results in adatom induced surface dipole. At the branch point: Charge Neutrality Level (CNL) of the virtual gap states of the complex band structure, the dipole moment becomes zero. The energy position of the adatom-induced surface state about the CNL depends on the electronegativity of the adatom. Adatoms, which are less electronegative with respect to the substrate, will become positively charged and assume acceptor character and lie above the CNL. Similarly, adatoms, which are more electronegative with respect to the substrate, will become negatively charged and assume donor character and lie below the CNL. For GaAs the branch point (CNL) of the Virtual Gap States (ViGS) lie 0.5 eV above the valence band maximum. More details of the gap states are described in Appendix I.

The interface dipoles formed by adsorption of saturated hydrocarbons and inert gas atoms (closed-shell systems) are explained by exchange like effects in the van der Waals interaction between adsorbate and the surface without any charge transfer (Lang, 1981; Bagus, *et al.*, 2002). Using the density functional theory (DFT), Michaelides, *et al.*, have shown that the position of the adsorbate above the surface has a significant effect on the surface potential ($\Delta\phi$), in addition to the adsorbate-substrate electronegativity difference (Michaelides, *et al.*, 2003).

1.7.1 Monovalent Adatom

The monovalent hydrogen and halogen atoms saturate the dangling bonds of elemental and compound semiconductors. An exposure of semiconductor surfaces to molecular halogens at room temperature leads to dissociative chemisorption. The atomic hydrogen chemisorption on GaAs surface, preferentially removes arsenic (As) (Petravic, *et al.*, 2003); this behaviour correlates with the dissociation energy of respective molecule. The chemisorption of molecules may occur either directly or via a weakly bound precursor state. The coverage of the surface by the adsorbates is usually measured in monolayers. One monolayer (ML) is defined as the total number σ_{hkl} of the substrate sites per unit area in respective bulk (*hkl*) planes. The Sticking coefficient $S(\theta)$ of a chemisorbed species can be defined as

$$\frac{d\theta}{d\theta_{imp}} = S_0 \left(1 - \frac{\theta}{\theta_s}\right)^n \qquad \ldots 1.5$$

S_0 is the initial sticking coefficient at zero coverage of adatoms; θ is the coverage measured in unit of Mono Layers (ML).

The adsorption induced surface reconstructions were monitored using LEED. The bond length of adatoms is determined using Surface extended X-ray absorption fine structure (SEXAFS). The position of adatoms with respect to the lattice of the substrate may also be determined using the technique of X-ray standing waves (XSW). The different adsorption sites on surfaces may directly be detected by temperature programmed desorption (TPD).

Adsorption of atomic hydrogen, molecular oxygen, sulfur, halides, alkali metals, and metals on elemental and compound semiconductors are studied under ultra high vacuum (UHV) conditions. The metal atoms are very mobile at semiconductor surfaces and tend to pile up as three-dimensional islands at room temperature (RT) and above.

Structural data on the adsorption of monovalent metal and non-metal atoms on semiconductor surfaces indicate the existence of covalent bond between adatoms and surface atoms of the substrate. A charge transfer between the adatoms and surface atoms due to their electronegativity difference leads to chemical shifts of the core level binding energies.

1.8 ADSORBATE-INDUCED WORK FUNCTION CHANGES

The charge separation between the adsorbed layer and the surface leads to a contribution to the dipole moment perpendicular to the surface. The electronegativity difference between the adatom and the surface atom induces electric dipole layer at the surface. Such a two-dimensional arrangement of dipoles (Figs. 1.5 and 1.6) may be described as an electric double layer or a dipole layer and the voltage drop across it leads to an increase or decrease of ionization energy of an initially clean surface depending on the direction of the respective dipole moment. The change in ionization energy does not grow linearly with the number of adsorbate atoms. Drawing the analogy with a parallel plate capacitor (Helmholtz equation for unionized monolayer), for adsorbate induced surface dipoles per unit area (N_{ad}) having normal component of dipole moment $p_\perp(\theta)$ the change in ionization energy (ΔI) can be written as (Taylor and Bayes; 1994).

$$\Delta I = \pm \left(\frac{e_0}{\varepsilon_0}\right) p_\perp(\theta) N_{ad} \qquad \ldots 1.6$$

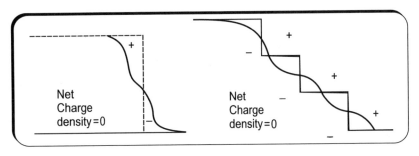

Fig. 1.5 Charge distribution at an atomically smooth and atomically stepped surface.

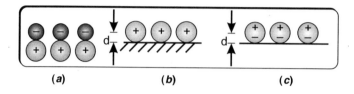

Fig. 1.6 Dipoles due to (a) covalent chemisorptions, (b) ionic chemisorptions and (c) physisorption on a metal surface. The charge center spacing is denoted as "d".

Dipole moment for square or hexagonal lattice as a function of coverage is given by

$$p_\perp(\theta) = \frac{p_\perp(0)}{\left(1 + 9\alpha_{ad} N_{ad}^{3/2}\right)} \qquad ...1.7$$

where $p_\perp(0)$ is the dipole moment of the isolated atom and α_{ad} is the polarizability of the adatom. By combining the above two equations,

$$\Delta I = \pm\left(\frac{e_0}{\varepsilon_0}\right)\frac{p_\perp(0)}{\left(1 + 9\alpha_{ad} N_{ad}^{3/2}\right)} N_{ad} \qquad ...1.8$$

The ionization energy $I = E_{vac} - E_{vs}$ can be determined from the photoemission experiments.

Work function of semiconductor surface can be written in terms of ionization energy as

$$\phi = I - e_0 V_s - (E_F - E_{vs}) \qquad ...1.9$$

Adatoms induce changes in the surface band bending ΔV_s in addition to a change in the ionization energy. The change in adsorbate induced work function can be written as

$$\Delta\phi = \Delta I - e_0 \Delta V_s \qquad ...1.10$$

The adsorbate-induced variation of the ionization energy may thus be obtained when changes of both work function and surface band bending are measured as a function of the coverage. The Kelvin probe operated in dark is used for monitoring the change in the work function during adsorption of adatoms on semiconductor surfaces. The surface band bending is determined using intensity dependant surface photovoltage (SPV), Photo Electron Spectroscopy (PES) and Electric Field Induced Raman Spectroscopy (EFIRS).

Alkali metals adsorbed on the elemental and compound semiconductor surfaces are always found to lower the ionization energy; it is expected from the difference in the electronegativity of the substrate atom and the adsorbate. Isolated metal adatoms on GaAs (110) surfaces introduces surface states of donor type and are attributed to covalent bonds between metal adatoms and surface gallium atom of the substrate. The Fermi level pinning position observed experimentally at low-coverage and low temperature shows a linear relationship with first ionization energies of the respective metal adatom.

Non-metallic adatoms like chlorine, sulfur and oxygen on GaAs exhibit surface states of acceptor character while most of the metallic atoms exhibit donor character. The adsorption of chlorine on cleaved GaAs (110) surfaces was found to increase the ionization energy upto 1.4 eV and to pin the Fermi level at 0.1 and 0.2 eV above the valence band maximum (VBM) on samples doped n and p-type, respectively. The adsorbate induced surface states are identified as ViGS of the complex band structure of GaAs (Appendix I). At enhanced exposure of chlorine, the Fermi level assumes a second pinning position at 0.55 eV above Valence band Maximum (VBM) for both types of doping. This change of pinning position is attributed to the formation of adsorbate-induced defects. More details of the metal induced gap states are given in Appendix I.

1.9 AN OUTLINE OF SURFACE CHARACTERIZATION TECHNIQUES

With the advent of new surface sensitive characterization techniques over the years, a detailed understanding of the physics and chemistry of surfaces has been developed. Many of these experimental techniques measure properties that are directly calculable using approximate theoretical methods. A combination of both experiment and theory is important in extracting quantities that are complicated by many physical effects.

The surface characterization techniques can be broadly classified into three groups depending upon the properties being studied: (*i*) surface structure, (*ii*) chemical composition and bonding and (*iii*) electronic and optical properties. Most of the experimental methods involve probing the surface with electrons, photons, ions, neutrons and then analyzing the elastically or inelastically scattered radiation from the valence band, core level, plasmons, and other defect related electronic structures. The bulk and surface-sensitive experimental conditions are achieved by varying the excitation energy, angle of incidence, or emission angle. A clean ultrahigh vacuum is one of the basic requirements for most of these techniques. High energy radiations used in these techniques may affect the surface *via* desorption, dissociation, oxidation, localized diffusion and ionization.

The Table 1.1 gives a feel for the surface and thin film analyses techniques; and Table 1.2 gives some of the popular microscopy techniques with their modes of operation used to probe the surfaces.

1.10 EXPERIMENTAL TECHNIQUES FOR SURFACE STRUCTURE EVALUATION

A knowledge of the atomic geometry at the surface plays a key role in determining the electronic properties. The experimental techniques such as low-energy electron diffraction (LEED), Reflection High Energy Electron Diffraction (REED), X-ray Standing Wave (XSW), Surface extended X-ray absorption fine structure (SEXAFS), X-ray photoelectron diffraction (XPD), and Scanning Tunneling Microscopy (STM) have successfully been applied to surface structure related properties. LEED is one of the oldest and most powerful techniques for the determination of surface structure (Many et al., 1965). The LEED technique employs electrons of energy about 150 eV (wave length close to inter atomic distance and low penetration depth) and the diffraction pattern reveals the surface structure. REED technique uses a high energy electron beam (10-100 keV) at grazing incidence in order to achieve surface sensitivity. The geometry of REED apparatus enables one to monitor online epitaxial growth in Molecular Beam Epitaxy (MBE). In adsorption studies, XSW and SEXAFS are being used to determine position of adsorbed atom with respect to the substrate lattice and bond length of adsorbed atoms, respectively. XSW has the advantage of being able to work under non-UHV conditions. The scanning tunneling microscopy (Binning et al., 1982) and scanning tunneling spectroscopy (Hamers et al., 1986) have opened up real space mapping of surface atomic and electronic structures. In XPD, the interference between the photoelectron wave reaching the detector and the wave scattered from the neighbouring atoms, give the local real-space environment of the emitting atom (Lee, et al., 1996).

1.11 CHEMICAL COMPOSITION AND BONDING

X-ray photoemission spectra (XPS) provides information principally on the localized core levels and energetic shifts (chemical shift) related to the chemical bonding and many-electron effects. Using soft X-rays, shallow core levels of specific atoms can be excited (low escape depth for emitted electrons) for chemical composition analysis of the surface. Auger electron spectroscopy (AES) is more sensitive for chemical composition analysis of the surface than XPS. Since XPS and AES cannot detect hydrogen, special techniques like nuclear-reaction analysis (NRA) and elastic recoil detection analysis are used. Synchrotron radiation as a high-intensity, tunable, polarized source of photons provides the relative contribution of photoelectron emission from different elements can be tuned to get optimum sensitivity (Brillson, 1982c).

1.12 ELECTRONIC AND OPTICAL PROPERTIES

Photoelectron spectroscopy is one of the widely used methods for studying electronic properties of bulk semiconductor and their surfaces. Using the various methods of Photoelectron energy distribution measurements, the angle integrated, angle resolved, and yield-type spectroscopy determination of both occupied and unoccupied two-dimensional surface electronic band structure can be made (Feuerbacher. *et al.*, 1978). Ultraviolet photoemission spectroscopy (Knapp and Lapeyre, 1976, Larsen *et al.*, 1981; Salmon and Rhodin, 1983) and photoemission threshold measurements (Sebenne *et al.*, 1975) are used for identifying the occupied surface states and structure of valence band. Unoccupied surface states are probed using inverse photoemission spectroscopy and electron energy-loss spectroscopy (Ludeke and Esaki, 1974).

The availability of the synchrotron radiation as a light source with a spectrum extending continuously from visible into X-ray region, which is highly polarized, has led to the development of new techniques in photoemission spectroscopy such as spin polarized photoemission spectroscopy (Feuerbacher *et al.*, 1978). Three-dimensional mapping of the band structure became easier due to the availability of continuous spectrum of photon energy. The highly polarized beam of synchrotron source has been used to determine adsorbate site and molecular orientation on ordered surfaces. The photoemission technique measures the work function and ionization energy of semiconductor surfaces under atmospheric conditions (Riken Keiki Co. Ltd.).

The surface optical effects produce reflection and absorption phenomena in the vicinity of the surface that is different from the bulk. The optical absorption experiments reveal the coupling effect between initial occupied and final empty states (Chiarotti *et al.*, 1971). The optical techniques such as ellipsometry, reflectance spectroscopy, surface photoconductivity and surface photovoltage contributed much to the understanding of the optical and electronic properties of the surface states (Lüth, 1975). The optical techniques such as Raman spectroscopy and photoluminescence can evaluate the surface electronic structure non-destructively.

The surfacephonons are studied experimentally by using high-resolution energy loss spectroscopy (HRELS) with low-energy helium atoms and electrons. Electric-Field Induced Raman Scattering (EFIRS) is employed to determine electric field strength in surface space charge layer and hence the surface band bending.

Nevertheless, much more work in terms of developing new techniques and exploiting the full potential of the existing technique is required to develop better characterization techniques specific to the surface studies.

Table 1.1 Techniques specific to the surface studies

	Technique	Incident beam	Emitted Beam	Spatial Resolution (µm)		Other features
				dia	thick	
EELS	Electron energy loss spectroscopy	Electrons	Electrons	0.01	0.05	Semiquantitative Valance States Non-Destructive
XPS	X-ray photoelectron spectroscopy	X-rays	Electrons	150	0.003	Surface structure
RHEED	Reflection high energy electron diffraction	Electrons	Electrons	100 Å	100 Å	Surface structure
LEED	Low energy electron diffraction	Electrons	Electrons	300 Å	3 Å	Surface /interface structure
TEM	Transmission electron microscopy	Electrons	Electrons	100 Å	2 Å	Surface topography Conductor
SEM	Scanning electron microscopy	Electrons	Electrons	20 Å	30 Å	Surface topography Conductor
STM	Scanning tunneling microscopy	Distance ($\approx$ 1-2 Å)	Tunneling ($\approx$ 1- 10Å)	3 Å	0.01 Å	Bias ($\approx 1mV$ to $\approx 1V$)

K.Wasa, M. Kitabatake, and H. Adachi, Thin Film Materials Technology, William Andrew, Inc. and Springer-Verlag GmbH & Co. KG, 2004

Table 1.2 Popular microscopy techniques with their modes of operation

Name(s) of technique	Acronym(s)	Mode of operation	What is measured?
Scanning Tunneling Microscopy	STM	Tunneling current controls z-regulating feedback loop.	Atomic-scale imaging of morphology (indirectly), or location of orbitals at particular energy levels, When tunneling voltage is varied, the measured current yields a spectrum. This variation of the technique is called scanning tunneling spectroscopy (STS) and yields information on both filled and empty states of the sample's band structure.
Atomic force microscopy or Scanning force microscopy	AFM, SFM	Cantilever-spring deflection controls z-regulating feedback loop.	Nanoscale measurements of surface morphology, materials properties, and forces between tip (which may be functionalized) and surface.
Friction force microscopy	FFM, LFM	Cantilever-spring deflection controls z-regulating feedback loop while torsional deflection of spring is displayed	Friction can be measured and differentiated on a nanometer scale. When this is related to the surface chemistry, this is often referred to as chemical force microscopy (CFM), and tips are often surface-treated in order to enhance contrast.
Magnetic force Microscopy	MFM	Deflection of cantilever spring caused by magnetic forces between magnetized tip and surface controls z-regulating feedback loop.	Magnetic field gradient above a sample.
Electric force microscopy	EFM	Deflection of cantilever spring caused by electrostatic forces between tip and surface controls z-regulating feedback loop.	Electric filed gradient or surface potential above a sample.

(Contd..)

Scanning Kelvin probe microscopy	SKPM	Capacitive force is measured between oscillating tip and surfaces while the sample voltage is varied until the electrostatic field is compensated.	Map of surface contact potential difference.
Scanning capacitance microscopy	SCM	Capacitance is measured between oscillating tip and surfaces while scanning a biased tip above the sample surface.	Map of surface capacitance.
Near-field scanning optical microscopy or Scanning near-field optical microscopy	NSOM, SNOM	An optical fiber with a small aperture is scanned in very close proximity to the sample, and the transmitted or reflected light is detected and of analyzed spectroscopically.	Optical image of a surface with resolution of −100 nm; optical images of smaller emitting species, such as single fluorescent molecule.

Various Scanning Probe Microscopy (SPM) Techniques. (After: E. Meyer, S.P. Jarvis and N.D. Spencer, MRS Bullentin, 29, 2004, p443).

1.13 KELVIN PROBE APPLIED TO SURFACE STUDIES

The Kelvin probe was first proposed by Lord Kelvin in 1898 to investigate the contact potential difference (CPD) between metallic surfaces (Kelvin, 1898). The Kelvin method of measuring the CPD has the advantage that it can be used for wide range of materials, temperatures and ambient. The unique feature of this technique is that the surface remains virgin even after the measurement; which is of critical importance in studying weakly bound adsorbates. The application of Kelvin probe method to study the electronic properties of semiconductor surfaces is emerging as a powerful tool (Shimizu *et al*, 2000; Amico *et al.*, 2000,). The Kelvin probe is used widely for studying the electronic properties of clean and intentionally modified surfaces and interfaces, including junctions: Schottky barriers and heterojunctions.

1.14 A BRIEF RECORD AND VERSATILITY OF THE KELVIN PROBE

Brattain and Bardeen employed Kelvin probe to study the electronic properties of etched and sand blasted germanium surfaces under various gaseous ambient (Brattain and Bardeen, 1953). Later, Allen and Gobeli used this technique for studying clean surfaces in ultra high vacuum along with other photoemission spectroscopy techniques (Allen and Gobeli, 1962). The Kelvin probe is extremely sensitive, <0.1 mV (Craig and Radeka, 1970) and has been operated over a wide range of temperatures and pressures (Lundgren and Kasemo, 1995). Further, this is an indirect, null-field method, *i.e.*, it does not run the risk of desrobing weakly bound adsorbates as in the case of photoelectric and field emission techniques. The Kelvin probe played a key role in the understanding the interfaces of homo- and hetero-junctions of various inorganic and organic semiconductors and Schottky barriers (Hayashi *et.al.*, 2002; Vilan *et al.*, 2000; Chan *et al.*, 2004). It is the basic tool in the surface photovoltage spectroscopy (SPS) technique to determine the energy levels of the occupied and empty surface states (Gatos and Lagowski, 1973) and surface band bending (Brillson and Kruger, 1981) in semiconductors. The surface potential measurements using Kelvin probe provides information about the polarization of adsorbed layer and equilibrium coverage of the adsorbed species. The Kelvin probe offers a fast, in-line, non-destructive method of detecting minute organic or metallic contaminants on wafer surface (Qcept Technologies). It also finds wide applications in electrochemistry (commonly referred as non-Nernstian potentiometry) for studying half-cell potentials and other charge transfer mechanisms in electrode materials both from liquid and gas phases (Janata and Josowicz, 1997; Badwal *et al.*, 2001). Also Kelvin probe is a useful tool for studying contact charging in organic thin-film insulators (Harris and Fiasson 1984; Manaka *et al.*, 2003; Nalbach and Kliem, 2000). The Kelvin probe and Kelvin probe force microscopy (KPFM) (Arakawa, *et al.*, 1997) facilitates CPD measurements in nanometer scale of spatial resolution. This list is only indicative. There is a vast amount of literature available where the Kelvin probe is used for many new and novel materials like organic semiconductor/metal junctions, Self-Assembled Monolayers (SAM) and biological surfaces.

■■■

Chapter 2

The Thermodynamics of Surfaces and Definition of Work Function and Measurement

2.1 INTRODUCTION TO THERMODYNAMICS OF CHARGES ON THE SURFACES

The study of work function of solid surfaces contributed much to the development of many devices and understanding of several phenomena. It is a common experience that the surface properties of metals and semiconductors are sensitive to the available surface energy and to the surrounding reactive species. Among the several techniques available to measure the work function, Kelvin's method of measuring contact potential difference (CPD) is more versatile and can be used for wide range of materials, temperatures and pressures. The concept of CPD is very important in explaining the theory of basic *p-n* junction and many surface properties; however, one does not find enough explanation in the common textbooks. CPD cannot be measured directly with any conventional voltmeter. In order to interpret the data of CPD by Kelvin probe, one needs a basic understanding of the thermodynamics of the surface.

This chapter presents a review of basic concepts and definition of chemical potential, work function and contact potential difference. The experimental techniques for measuring the work function (diode and field emission techniques) are outlined. For an easy understanding, the ensemble approach of the statistical thermodynamics is followed.

The removal of an electron from solids, both bulk and surfaces, involves complex processes. In general, for semiconductors, the work function is defined as the energy required to remove the electron from the Fermi level to the vacuum level. The Fermi level of the surface (even without any excitation energy) is different from the bulk. It may not be possible to define the Fermi level for all the materials. How does one then define the true work function of the material? The definition of true work function, a macroscopic property, depends upon many microscopic processes; hence, one invokes assumptions regarding specific models to describe the microscopic constituents of the system. While defining the work function, the Fermi level is often equated with chemical potential. The concept of (electro) chemical potential for a thermodynamical system containing many particles in a fluid can be conveniently adapted for defining the work function without invoking any specific model. A solid phase (metal or semiconductor) always constitutes immobile neutral atoms, ions and mobile charge carriers: electrons and holes. It is often convenient to consider the electrochemical potential of electrons in its equilibrium fluid phase (free electrons) in a solid for the thermodynamical treatment.

2.2 CHEMICAL POTENTIAL AND CONTACT POTENTIAL DIFFERENCE

2.2.1 Surface Energy

(All the symbols used in this chapter are common to the stastatical thermodynamics treatment followed in the standard texts: Reif and Tolman).

Consider a homogeneous system with N number of particles (for example, electrons on the surface of a solid) at a temperature T, pressure p and occupying a volume V with an internal energy U and entropy S. If the number of particles is allowed to vary, either by chemical reaction or transfer of matter across the boundaries, the basic thermodynamic relation $dU = TdS - pdV$ may be expressed in the form (Reif, 1965)

$$dU = TdS - pdV + \sum_{i=1}^{m} \mu_i dN_i \qquad \ldots 2.1$$

where m is the total number of states and μ_i is the chemical potential per molecule and has the form

$$\mu_i \equiv -T \left(\frac{\partial S}{\partial N_i} \right)_{U,V,N} \qquad \ldots 2.2$$

Upon differentiation and using the definition of dU, one obtains the Gibbs equation:

$$SdT - VdP + Nd\mu = 0 \qquad \ldots 2.3$$

The surface has to be considered differently from the bulk. For instance, the pressure in the bulk of an isotropic solid is equal in all the directions; whereas, the pressure on the surface plane is highly anisotropic. To create a surface, *e.g.*, by cleaving, one has to spend energy that is proportional to the additional surface area A created,

$$U = TS - PV + \mu N + \gamma A \qquad \ldots 2.4$$

The proportionality factor γ is the *surface tension*

$$\gamma = dU / dA \qquad ...2.5$$

The surface tension can be defined as the reversible work of formation of a unit area of surface at constant T, V and μ for a single component system. The surface tension is the two-dimensional analog to the pressure.

However, in the case of the work done for a volume, pdV, the work to increase the volume, where p is always normal to the surface. In the case of increasing the area, keeping the volume constant, the surface tension is always parallel to the surface. The unit of γ is force per unit length (Newton/meter).

The increase of surface area can be achieved by *stretching*. One can calculate the increase in the energy with the conventional theory of elasticity:

$$dU = TdS - VdP + Nd\mu + A\sum_{i,j}\sigma_{ij}d\varepsilon_{ij} = 0 \qquad ...2.6$$

where σ_{ij} and ε_{ij} are the components of the surface stress (force per unit length) and strain (deformation) tensors respectively along the i direction, where j denotes the surface normal. Using $dA/A = Sd\varepsilon_{ij}\delta_{ij}$, the corresponding Gibbs equation, taking into account the surface is:

$$Ad\gamma + Sdt - VdP + Nd\mu + A\sum_{i.j.}(\gamma\delta_{ij} - \sigma_{ij})d\varepsilon_{ij} = 0 \qquad ...2.7$$

Now, one can separate the thermodynamic quantities in the bulk and at the surface part (*e.g.*, $S = S_b + S_s$, etc.). Applying the Gibbs-Duhem relation for the bulk, it can be shown that:

$$S_s = -A\left[\frac{\partial \gamma}{\partial T}\right] \qquad ...2.8$$

That is, the specific surface entropy is given by the temperature dependence of γ. Typically, $\gamma = \gamma_0(1-T/T_c)^n$ (van der Waals-Guggenheim semi-empirical relation) with $n \approx 1$ for metals, where T_c is the critical T (at which the solid phase vanishes). Also:

$$\sigma_{ij} = \gamma\delta_{ij} + \left[\frac{\partial \gamma}{\partial \varepsilon}\right]_T \qquad ...2.9$$

Note that in a solid, *the surface tension and surface stress are not identical*. In contrast, in a liquid, the surface tension is independent of small strains, since the liquid adapts to perturbations.

The minimum surface energy determines the morphology and composition of surfaces and interfaces. The minimization of energy leads to a spherical equilibrium shape in an isotropic liquid (in the absence of gravity). In crystalline solids, the surface tension depends on the crystal plane and direction. Thus, the equilibrium shape of a crystal is not obtained by minimizing the surface area alone but the integral $\int \gamma(n)dA$, where n is an orientation vector. Because γ is low for some crystal planes, faceting is energetically favoured, even if it implies a larger surface area.

It is important to note that the formation of the equilibrium shape requires sufficient mobility (or fast kinetics), not just the thermodynamics. The equilibrium shapes can be calculated but it is easier to use a graphics method, the *Wulff construction*: the surface tension is plotted in polar coordinates vs. the angle with respect to a particular direction. The minimization mentioned above implies constructing the surface from the inner envelope of planes perpendicular to the radius vector.

2.2.2 Chemical Potential

There are a number of ways one can understand the chemical potential; this section presents a summary.

Chemical potential of a system measures the change in its internal energy due to an increase or decrease in its chemical constituents; it has a special significance in chemical thermodynamics.

Chemical potential is generally interpreted as the tendency (and it is measure) of particles to diffuse from a higher potential to a lower chemical potential (Baierlein, 2001). The chemical potential can be written in many forms (based on the variables describing the system):

$$\mu_i \equiv \left(\frac{\partial E}{\partial N_j}\right)_{S,V,N} \equiv \left(\frac{\partial F}{\partial N_j}\right)_{T,V,N} \equiv \left(\frac{\partial G}{\partial N_j}\right)_{T,p,N} \qquad ...2.10$$

The chemical potential is often referred to as Gibbs free energy per particle ($\mu = G/N$) when only one particle species is present.

When two systems are placed in close thermal contact with one another, the particles move (number of particles and the energy) from one system to the other till such time the thermal and mechanical equilibrium is established. The state of equilibrium is said to be established when the chemical potential attains identical value in both the systems. At equilibrium, there is no net flow of particles and energy.

For two systems in diffusive and thermal contact, the entropy will be a maximum with respect to the transfer of particles as well as to the transfer of energy. The quantity, which governs the new equality, is defined as chemical potential, μ.

$$\frac{-\mu}{\tau} = \left(\frac{\partial S}{\partial N}\right)_{U,V} \qquad ...2.11$$

$\tau = k_B T$, T is temperature. Chemical potential keeps the number of particles constant in the system at a finite temperature (T).

For N electrons in a volume V (electron density, $n = N/V$), in the limit of large system, the Helmholtz free energy F per unit volume approaches a smooth function f of number density and temperature.

$$\lim_{n,v \to \infty} \frac{1}{V} F(N,V,T) = f(N,T) \qquad ...2.12$$

for large N and V,

$$F(N,V,T) = Vf(n,T) \qquad ...2.13$$

Chemical potential is defined as the change in free energy while adding one particle to the system, keeping T and V fixed,

$$\mu = F(N+1,V,T) - F(N,V,T) \qquad ...2.14$$

Allocation of the total energy that maximizes the number of accessible states is the law of increase of entropy.

In the independent particle approximation (negligible mutual interaction), calculation of chemical potential requires the evaluation of partition function (Z) using classical or quantum statistical mechanics. The partition function can be chosen from proper ensembles. In Canonical ensemble

$$Z = \sum_v e^{-\beta E_v} \qquad ...2.15$$

the sum is overall possible states v of the particles constituting the system and $\beta = \dfrac{1}{k_B T}$. The Helmholtz free energy F can be expressed in terms of the partition function in the form (Reif, 1965)

$$F \equiv U - TS = -k_B T \ln Z \qquad ...2.16$$

When there is an exchange of partices and energy between the two systems, the partition function of the grand canonical ensemble is most appropriate to describe the system (Tolman, 1938). The grand canonical ensemble gives

$$\Omega = U - TS - \mu \hat{N} = -k_B T \ln Z, Z = tre^{-(\beta H - \mu \hat{N})} \qquad ...2.17$$

Trace (tr) involves summing over energy states with various numbers of particles. Advantages of using the grand canonical ensemble are discussed by Kaplan (Kaplan, 2004). The total internal energy and the total number of particles in grand canonical ensemble are given by

$$U = \frac{\sum_{v,N} E_v(N) e^{-\beta[E_v(N)-\mu N]}}{\sum_{v,N} e^{-\beta[E_v(N)-\mu N]}} \text{ and } \hat{N} = \frac{\sum_{v,N} N e^{-\beta[E_v(N)-\mu N]}}{\sum_{v,N} e^{-\beta[E_v(N)-\mu N]}} \quad ...2.18$$

$$F = \Omega + \mu \hat{N} \text{ and } \mu = \frac{\partial F}{\partial \hat{N}} \quad ...2.19$$

For multi component system ($i, j,....$), $\mu \hat{N}$ is replaced by $\mu_i \hat{N}_i$ and the derivative is taken keeping the other $\hat{N}_j$ constant.

For a system having many phases (a, b, c....) in equilibrium, the chemical potential of each species (electrons, holes, ions and other chemical species) in the system should be the same in all phases (Zemansky, *et al.*, 1975).

$$\mu_i^a = \mu_i^b = \mu_i^c = ... \quad ...2.20$$

This equilibrium requirement is the one that forms the basis for explaining the junction behaviour of semiconductor devices. Lack of thermodynamic equilibrium has been reported for wide band gap semiconductors (nearly insulating) and metal induced surface electronic states (Koch *et al.*, 2003).

Thermodynamic limit is not required for the validity of chemical potential obtained using grand canonical ensemble; it is exact even for finite systems like quantum dots (Kaplan, 2004). The zero-temperature limit μ can be accounted better in grand canonical ensemble (Kaplan, 2004).

Evaluation of partition function for real systems requires many approximations and advanced molecular dynamics numerical simulation procedures (Lyubartsev, *et al.*, 1992). The chemical potential plays a key role in evaluating the chemical-equilibrium constants of any chemical reaction. For example, when a battery is in open circuit, the conduction electrons in the two metal terminals generally have different chemical potential. The potential difference can be related to the chemical potential and stoichiometric coefficients of the particles whose reaction provides power to the battery (Baierlein, 2001). The chemical potential of photons is zero (Baierlein, 2001).

2.3 CHEMICAL POTENTIAL OF FEW-ELECTRON SYSTEM OR ATOMIC CHEMICAL POTENTIAL

The concept of chemical potential for an isolated microscopic quantum mechanical system can be formulated from the density functional theory (DFT) (Katriel, *et al.*, 1981). Chemical potential is the derivative of ground state energy, $E(N)$ with respect to number of electrons, N at constant potential, v.

$$\mu = \left(\frac{\partial E(N)}{\partial N}\right)_v \qquad ...2.21$$

For neutral atoms, viewed as a one atom thermodynamic system representative of grand canonical ensemble, the chemical potential can be approximated by, $\mu = -\frac{(I_1 + A_1)}{2}$, in which I_1 and A_1 are the atom's first ionization potential and electron affinity, respectively (Gyftopoulos, and Hatsopoulos, 1968). The concept of electronegativity (χ) and chemical potential are equivalent (Parr *et al.*, 1978). Electronegativity (tendency to attract electrons) is the negative of chemical potential (tendency to escape),

$$\chi = -\mu = \frac{I + A}{2} \qquad ...2.22$$

This is the familiar Mulliken's definition of electronegativity of a neutral atom at 0K (Mulliken, 1934). The valence state electronegativity differences drives charge transfers on molecule formation. For a many electron system the chemical potential,

$$\mu = -\frac{(I + A)}{2} \qquad ...2.23$$

where I and A are the first ionization potential and electron affinity, respectively (Kaplan, 1973). Using density functional theory, Politzer *et al.*, found good correlation between atomic chemical potential and electrostatic potential created by the nuclear and electronic charge at a radial distance from the nucleus (Politzer *et al.*, 1983).

2.4 FERMI ENERGY AND CHEMICAL POTENTIAL IN SEMICONDUCTORS

(The aim of this section is to explain the conditions at which the Chemical potential can be equated with Fermi energy).

The chemical potential plays a more fundamental role in the formulation of Fermi-Dirac statistical description of free electron gas. For a system with N electrons and having characteristic energy, $k_B T$, the probability of occupancy of one electron state E_i depends on the difference $E_i - \mu$,

$$f_i^N = \frac{1}{e^{(E_i - \mu)/k_B T} + 1} \qquad ...2.24$$

where k_B is the Boltzmann constant. The difference ($E_i - \mu$) is the total entropy change when one particle is added to the single particle energy state E_i (Baierlein, 2001). At $T = 0$ all states of lowest energy are filled up to m and all states above m are empty. This energy is called the Fermi energy. For temperature other than 0K, the Fermi energy differs slightly from μ.

At finite temperature T, if $E_i = \mu$, then $f_i^N = \dfrac{1}{2}$,

In addition to the thermodynamical concept, the quantum mechanical aspects of the system are of great importance in quantifying the work function from the microscopic details of the various interactions.

For systems of interacting particles, as in the case of electrons in real solids, the calculation of chemical potential is more complex.

The quantity $\mu(T)$ is usually determined in such a way that the total number or the number density of particles comes out correctly. For any set of non-interacting particles, n can be determined as

$$n(\mu,T) = \int_{-\infty}^{\infty} f(E,\mu,T)\rho(E)dE \qquad \ldots 2.25$$

where $\rho(E)$ is the density of energy levels. If n and $\rho(E)$ are known, the above equation can be solved for $\mu(T)$.

The Sommerfeld expansion for chemical potential of a three-dimensional, non-interacting Fermi gas is related to the Fermi energy by the equation

$$\mu(T) = E_F\left[1 - \frac{\pi^2}{12}\left(\frac{k_B T}{E_F}\right) + \frac{\pi^4}{80}\left(\frac{k_B T}{E_F}\right)^4 + \ldots\right] \qquad \ldots 2.26$$

where E_F is the Fermi energy, k_B is the Boltzmann constant. Hence, the chemical potential is approximately equal to the Fermi energy at temperatures of much lower than the characteristic temperature of the Fermi energ (E_F/k_B). The characteristic temperature is of the order of 10^5 K for metals, hence at room temperature (300 K), the Fermi energy and chemical potential are essentially equivalent. This is significant since it is the chemical potential, not the Fermi energy, which appears in Fermi-Dirac statistics. Strictly, $\mu(T) = E_F$ is true only for $T = 0$. A knowledge of $\mu(T)$ is very important in explaining the temperature dependence of transport properties of real solids (Villagonzalo et al., 1999).

The chemical potential μ of an intrinsic semiconductor is obtained in the canonical ensemble by equating the total number of electrons in the conduction band with the total number of holes in the valence band. The standard treatment leads to,

$$\mu = E_C - \frac{1}{2}E_g + \frac{3}{4}k_B T \ln\left(\frac{m_V}{m_C}\right) \qquad \ldots 2.27$$

It is assumed that, $E_g, (E_C - \mu), (\mu - E_V) \gg k_B T$

where E_C is the energy at the bottom of the conduction band, E_g is the band gap energy, m_V and m_C are the effective masses for the valence band and conduction band, respectively. The Fermi-Dirac distribution function is not valid in the canonical ensemble for an intrinsic semiconductor unless the number of carriers in the conduction band (N_{CB}) is large. At sufficiently low temperature, that is when the number of electrons in the CB is low, the μ starts to swing towards the CB and show appreciable volume dependence (Shegelski,1986). But the standard result for μ given by equation 2.27 indicates that μ goes to the middle of the gap as T goes to zero.

Within the usual non-interacting electron model of an intrinsic semiconductor with N electrons, the chemical potential can be expressed as,

$$\mu(N) = \frac{1}{2}\left[E(N+1) - E(N-1)\right] \qquad \ldots 2.28$$

The above expression can be written in terms of ionization energy $I(N)$ and electron affinity $\chi(N)$,

$I(N) = E(N-1) - E(N), \ \chi(N) = E(N) - E(N+1)$ as,

$$\mu(N) = -\frac{1}{2}\left[I(N) + A(N)\right] \qquad \ldots 2.29$$

the behaviour of chemical potential as a function of T differs appreciably at very low temperature. At 0K, $\mu(N) = E_C$ but the thermodynamic limit ($V, N \to \infty$, N/V fixed) is $\mu(N) \to (E_C + E_V)/2$ as $T \to 0K$.

For semiconductors with excess donors $\mu \to$ donor level as $T \to 0$

At the zero temperature limit, $\mu(N,T,V)$ goes from E_V to E_C, as N goes from $N_{VB} - 1$ to $N_{VB} + 1$, here N_{VB} is the number of electrons in the completely occupied valence band at $T = 0$.

Interestingly, the semiconductors have no Fermi surface.

In a semiconductor there are few electrons in the conduction band, which can be considered as a non-degenerate gas. The exchange and correlation energies are small (Fan.,1942).

2.5 ELECTROCHEMICAL POTENTIAL

The electrochemical potential $\bar{\mu}_i^\beta$ of the component i (an electron, a hole, or an ion) in an isotropic, homogenous, chemical phase β with total volume V, pressure P, temperature T, and entropy S is defined for a system in thermostatic equilibrium as

$$\bar{\mu}_i^\beta \equiv \left(\frac{\partial G}{\partial N_i}\right)_{P,T} \equiv \left(\frac{\partial H}{\partial N_i}\right)_{S,P} \equiv \left(\frac{\partial A}{\partial N_i}\right)_{T,V} \equiv \left(\frac{\partial U}{\partial N_i}\right)_{S,V} \quad \ldots 2.30$$

The quantities G, H, A, and U are, respectively, the total Gibbs energy, the total enthalpy, the total Helmholtz energy, and the total internal energy of the entire system, and N_i is the number of the ith component contained in the system.

The electrochemical potential $\bar{\mu}_i^\beta$ separable into a chemical potential μ_i^β and an electrical potential ϕ^β, in the form

$$\bar{\mu}_i^\beta = \mu_i^\beta + q\varphi^\beta \quad \ldots 2.31$$

where q is the charge on the component i. The chemical potential depends on the chemical or atomic nature of the phase "β", and independent of surface and external conditions. The electrical component $q_i \phi^\beta$ is determined by the distribution of electrical charges in the whole system as well as in the surroundings such as a deposited dipole layer.

2.6 ELECTROCHEMICAL POTENTIAL DISTRIBUTION WITH ELECTRICAL CURRENT AND TEMPERATURE DISTRIBUTION

The thermodynamic theory establishes linear relations between certain "flows" and the generalized "driving force" which causes these flows. Special cases of such relations are the Ohm's law in an isothermal phase and Fourier's law of heat flow across a phase without electrical current. The local electrical potential gradient in terms of local electrical current density J, temperature gradient ∇T, and chemical potential gradient $\nabla \phi$ can be written as

$$\nabla \phi = \left(\frac{1}{q}\right)\nabla \mu + \left(\frac{1}{q}\right) S^* \nabla T - \left(\frac{1}{\sigma}\right) J \quad \ldots 2.32$$

where σ is the electrical conductivity and S^* is entropy flow per particle. A large deviation in the carrier concentration in any region should not differ appreciably from the distribution which would obtain in the same region in the absence of a current.

2.7 DEFINITION OF WORK FUNCTION

2.7.1 Work Function Based on Electrochemical Potential

The true work function $q\Phi$ of a uniform surface of an electronic conductor is defined as the difference between the electrochemical potential μ of the electrons just inside the conductor and the electrostatic potential energy – $e\Phi_a$ of an electron just outside (in the vacuum). Thus,

$$\Phi = -\Phi_a - \left(\frac{\bar{\mu}}{e}\right) = \Phi_c - \Phi_a - \left(\frac{\mu}{e}\right) \qquad ...2.33$$

where Φ_c is the electrostatic potential inside the conductor.

The first two terms on the right of the above equation represent the potential difference between the inside and outside of the conductor and depend on the condition of the surface. The chemical potential, μ on the other hand, is a volume property and is sometimes referred to as the inner work function.

The electrostatic potential energy of an electron just outside a metal surface is dependent on the distance from the surface. In the absence of external electric fields, the potential experienced by the electron is given by the image potential (assuming the surface as an equipotential plane):

$$V(r) = -\frac{e}{16\pi\varepsilon_0 r} \qquad ...2.34$$

This potential which corresponds to ϕ_a in equation (2.33) is a function of distance r. At distances of few angstroms from the surface, the image potential form given above is not valid (Cutler and Davis, 1964). In practice, the potential does not change significantly for distances greater than 10^{-5} cm from the surface. As shown in Fig. 2.1, $\phi_a \to 0$ as $r \to \infty$ and the work function may be defined using the value of ϕ_a at ∞ (Knapp, 1973).

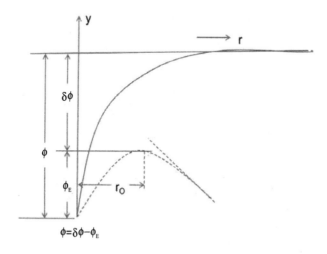

Fig. 2.1 Image potential of a charge with the distance from the nucleus (Surface).

The presence of accelerating electric field E reduces the work function by an amount given by:

$$\delta\varphi = \left(\frac{e}{4\pi\varepsilon_0}\right)^{\frac{1}{2}} E^{\frac{1}{2}} \qquad ...2.35$$

This reduction in work function is called Schottky effect. It plays an important role in experiments involving electric high fields such as field emission.

2.7.2 Work Function Based on the Contact Potential Difference

When two conductors 1, 2, are kept at the same temperature and connected electrically through a circuit containing no source of e.m.f., electrons will flow from one conductor to the other until an equilibrium state is reached. In this equilibrium state, the electrochemical potential of the two conductors must be equal. The equation 2.33 implies that there must be a difference of potential between a point just outside conductor 1 and a point just outside conductor 2, given by the difference of the two work functions:

$$\Phi_{a1} - \Phi_{a2} = \Phi_2 - \Phi_1 \qquad ...2.36$$

This is called the contact potential difference (CPD) between 1 and 2; the existence of such a difference is called Volta effect.

2.7.3 Work Function Based on Thermionic/Field Emission

A condition for (thermionic) emission is that the electron energy (perpendicular to the surface) be greater than the height of the potential barrier at the surface:

$$\frac{1}{2}mv^2 > \phi \qquad ...2.37$$

Some of the electrons situated near the surface will escape with maximum retained energy; these electrons must surmount a potential step at the surface. The height of this step ϕ is the work function.

2.7.4 Note on the Diversity of Work Function

There are several definitions for the work function available in the literature. Also, for the same material, there are different work functions quoted in the literature (Table 2.1). The diversity may be because of the methods of measurements: thermionic-(or field) emission and photoemission techniques.

The equations 2.33, 2.36 and 2.37 provide three definitions for the work function. A common feature of the above definitions is that they all refer to specific electron energies at specific sites at the surface of the emitter material. The resultant work function value will be severely biased towards the patches with the lowest work function, even if these cover only a very small fraction of the total surface area. Suitable corrections must be applied to the calculated values due to thermionic Schottky barrier lowering, variation of work function with temperature and linearity of the electric field in the vicinity of the emitter etc (Riviere and Myhra, 1988).

2.8 WORK FUNCTION : BAND MODEL

The potential of an electron inside a crystalline solid is periodic with the periodicity of the lattice (Kronig–Penny model). The potential increases at the "ideal surface" forming a barrier, which keeps the electrons confined to the solid. The potential barrier in real solid surfaces is influenced by several factors: for example, the charges inside the bulk and on the surface. The asymmetric distribution of electrons around the positive charges at the surface results in a double layer, which produces a potential difference across the surface. The work function is defined as the amount of energy required to remove an electron from the Fermi level to the vacuum level. Fig. 2.2 shows the band model of the solid. The position of the Fermi level depends upon the details of the electron band structure, consequently, the surface work function is dependant on the modified or modulated surface.

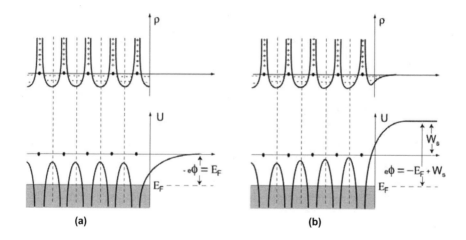

Fig. 2.2 (a) Undistorted infinite crystals, (b) Distorted infinite crystal (surface dipole).

2.9 THE VACUUM LEVEL

The measurement of work function, ionization energy and electron affinity of the solids are measured with reference to the vacuum level. The energy of an electron at rest in vacuum at infinite distance from the surface of the solid: $E_{VAC}(\infty)$, is generally taken as constant and as the reference; this assumption is valid when the charge distribution at the surface does not differ from the bulk. The charge distribution of a real surface is significantly different from the bulk resulting in the formation of dipole layer. In order to account for the influence of the surface, the vacuum level is defined with respect to the electron at rest just outside the surface $E_{VAC}(S)$ as shown in Fig. 2.3. For a dipole layer of finite extension, with a representative length L, the potential energy of an electron at certain range of distance is independent of x. For $x \gg L$, the dipole layer can be regarded as a point dipole, and the potential decreases as x^{-2} (Slater and Frank, 1947). Thus, the contribution of the surface is included by taking into account the flat portion of the potential energy curve (Fig. 2.2). The degree of tailing depends on the surface. The dependence of the work function on the surface can be ascribed to the difference in the tailing of the electron cloud at different locations.

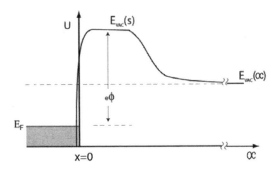

Fig. 2.3 The energy level diagram near the surface modified by a surface dipole layer.

2.10 NEGATIVE WORK FUNCTION

The thermalized positron emitted either as positron or positronium atom from most of the metal surfaces is closely related to the electron work function of the metal. Since energy is emitted when positron or positronium atom leaves the metal surface, it is known as "negative work function". For positive work function values of positronium, no positronium or positron emission will be produced from thermalized positrons.

Similar to the expression for the electron work function, positron work function can be written as

$$\Phi^p = \Delta\Phi^p - \mu^p \qquad ...2.38$$

where $\Delta\phi^p$ is the lowering of the mean electrostatic potential at the surface and μ^p is the chemical potential of positron in the material. The mean density of positron is much less than the electron density generally present on the solid surface.

Also, the positron work function can be written as,

$$\Phi_+ = -\mu_+ - \mu_- - \Phi_- \qquad ...2.39$$

where μ_+ is positron chemical potential, μ_- is electron chemical potential and Φ_- is electron work function.

Work function of metals and their electronegativities are found to be linearly related.

2.11 SURFACE AND BULK WORK FUNCTION OF METALS AND SEMICONDUCTORS

For work function and Contact Potential Difference (CPD) measurements, several techniques such as calorimetry, thermionic emission, field emission, photoemission, vibrating capacitor (Kelvin probe) and retarding-field electron-beam method, etc are available.

The experimental methods for measuring the work function may be classified into two groups: absolute and relative. Absolute measurements generally employ electron emission from the sample induced by photon absorption (photoemission), thermal energy (thermionic emission) or due to an electric field (field emission). All relative methods make use of the electric field created between the sample and the reference electrode.

For the sake of completeness, three techniques to determine the work function: diode method, field emission technique and photo-electric method are outlined here. The work function for several metals measured by different techniques (compiled from the literature) is given in Table 2.1.

2.11.1 Work Function Measurements Based Upon the Diode Method

The diode method is a widely used technique for measuring the surface potentials of gases adsorbed on metal surfaces (Knapp, 1973). This technique is based on the principle that the anode current (more precisely, the current –voltage characteristics) of a thermionic diode operates in two modes: the retarding field and the space charge limited mode, depending upon the work function of the anode. In the retarding field mode, the anode is negative and in the space charge limited mode, it is slightly positive with respect to the cathode. This approach is valid for anodes having uniform work function (single crystal). The current flowing through the anode may be expressed in the form

$$I_a = f(\phi_a - V_a) \qquad ...2.40$$

Where I_a is the anode current, V_a is the applied anode voltage, ϕ_a is the anode work function and f is a monotonic function. Now, if I_a is plotted against V_a to give the anode current-voltage characteristic, the curves for retarding field and space charge limited modes, can be represented by

$$I_a = f(\phi_a - V_a) \qquad ...2.41$$

$$I_a^* = f(\phi_a - V_a^*) \qquad ...2.42$$

And if the function f does not change during the adsorption process at the diode anode, two parallel curves will be obtained. The separation between these curves in the V_a direction is then given by $\Delta V = V_a - V_a^* = \phi_a - \phi_a^*$

It is more practical experimentally to hold I_a constant during the adsorption process and measure the change in V_a. Generally, it is necessary to check the current-voltage characteristics for parallelism before and after the adsorption process.

If the anode has non-uniform areas (known as patches, explained in Chapter 2.14), the diode methods will give true mean surface potentials if the patch size of the anode is small compared with the anode-cathode separation. It is because the electrostatic potential outside the surface attains a constant value ϕ_0 at distances from the surface which is large compared with the patch dimensions. Hence, in the diode method, the electrons will 'see' only the mean work function of the surface rather than the individual patch work functions.

The versatility of the diode method lies in its essential simplicity and the absence of a reference electrode and electrode contamination problem. The main limitation of the method is that it should be operated at low pressures; thus it restricts the types of adsorbates. The other restriction is that the adsorbate should not interact chemically with the filament. (More details can be seen in: Modern Techniques of Surface Science By D. P. Woodruff, T. A. Delchar, Cambridge University Press, 1994)

The Schottky diode method also provides a simple way to estimate the work function of metals. The metal–semiconductor (MS) Schottky diode is formed by deposition of the metal on the semiconductor. The diode equation contained the work function; the detailed procedure is explained in the Patent (United States Patent 7045384, 2006). The work function of tin doped indium oxide (ITO) has been estimated using this technique (Balasubramanian and Subrahmanyam, 1991).

2.11.2 Field Emission Technique of Measuring Work Function

The basic principle of the method is that the tunneling electron current in an emitter – collector geometry in vacuum is dependent on the work function of the emitter tip. The method makes use of the fact that field-emitted electrons are almost monoenergetic and have a well-defined energy (Vorburger *et al*, 1975). This energy enables one to determine the height of a potential barrier which the electrons have to surmount in a retarding field. The potential barrier is governed by the work function of the electron collector.

The Fowler-Nordheim equation describing the tunneling of electrons through the emitter tip may be written in the form

$$I/V^2 = a \, exp \, (-b \, \varphi^{3/2} / cV) \qquad \ldots 2.43$$

where a, b and c are constants. When ln I/V^2 is plotted against $1/V$, it yields a straight line with the slope proportional to $\varphi^{3/2}$. Thus, successive measurements on an emitter, first clean and then gas covered, provide two plots whose slopes will be in the ratio of the work function to the power $\frac{3}{2}$. In this method, the average work function of the clean emitter is to be ascertained before the work function of the adsorbed surface can be determined. If the field $F (= cV)$ is known, then the absolute work function can be determined. Experimentally, it is convenient to use a retarding potential analyzer, which measures the total energy distribution of the field-emitted electrons.

For spectroscopically pure gold and nickel (under ultra-high vacuum conditions) the work functions are measured by the field emission technique as 5.45 and 5.22–5.27 eV with an accuracy of ±0.01 eV (Holscher, 1966).

2.11.3 Photoelectric Method of Measuring Work Function

Photoelectric effect is: photons of suitable wavelength when incident on a metal surface, electrons are emitted ($h\nu > e\phi$, where ϕ is the metal work function). A theoretical analysis (Fowler, 1931) shows that the quantum yield I (photoelectrons per light quantum absorbed) is related to ν by the equation

$$I = bT^2 F(\mu) \qquad ...2.44$$

where b is a constant for the small range of frequencies near to the photoelectric threshold. F is expressible as a series in μ, where $\mu = (h\nu - e\phi)/k_B T$.

In practice, it is convenient to rearrange equation 2.44 in the form

$$\ln(I/T^2) - \ln b = \ln[F(h\nu - e\varphi)/k_B T] \qquad ...2.45$$

Then two plots are required: first, $\ln(I/T^2)$ against $h\nu/k_B T$, based on the experimental observations, and secondly $\ln F(h\nu/k_B T)$ against $h\nu/k_B T$. The values of function F can be found from the tabulated results, and $e\phi/k_B T$ is a constant for a given surface. The horizontal displacement which is necessary to superimpose these plots is $e\phi/k_B T$; the vertical displacement is $\ln b$.

In the region which is not too close to the threshold frequency ν_0 defined by $h\nu_0 = e\phi$, the equation 2.44 can be approximate equation as

$$I = bh^2(\nu - \nu_0)^2 / 2k_B^2 \qquad ...2.46$$

It is now possible, and often more convenient, to plot $I^{\frac{1}{2}}$ against ν to obtain b and ν_0. This is effectively a zero temperature approximation.

2.12 CONTACT POTENTIAL DIFFERENCE: METAL–METAL SYSTEM

The contact potential difference between two electronic conductors in thermostatic equilibrium can be obtained from the concept of equilibration of the electrochemical potential of electrons in it. Consider two homogeneous ideal metals X and R having different work functions and maintained at the same temperature (thermodynamic equilibrium) in charge free space (Fig. 2.4).

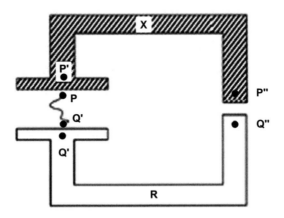

Fig. 2.4 Schematic arrangement for defining the isothermal Volta potential difference and the true work function.

The electrochemical potential of system X and R before contact is given by

$$\bar{\mu}_i^X = \mu_i^X - e\Phi_i^X \qquad \ldots 2.47$$

$$\bar{\mu}_i^R = \mu_i^R - e\Phi_i^R \qquad \ldots 2.48$$

'i' refers to the initial states

The true work function $_eW^X$ of a uniform metal surface X is defined as the difference between the electrical potential energy $-e\Phi_i^v$ of an electron in the vacuum just outside the surface of X and the electro-chemical potential, $\bar{\mu}_i^X$ of an electron just inside the metal.

$$_eW^X = -e\Phi_i^v - \bar{\mu}_i^X \qquad \ldots 2.49$$

$$_eW^R = -e\Phi_i^V - \bar{\mu}_i^R \qquad \ldots 2.50$$

Let Φ^{vX} and Φ^{vR} be the electrical potentials at point P just outside the surface X and at point Q just outside the surface R, respectively.

After making the contact between X and R, $\bar{\mu}_i^X$ and $\bar{\mu}_i^R$ adjust themselves by loosing or gaining electrons until the condition for equilibrium is satisfied. In metals the density of conduction electrons is very high and hence the space charge created by electron transfer is very narrow at the contact surface. The work function of the metals is not changed by the contact (Fan, 1942).

$$\bar{\mu}^X = \mu^X - e\Phi^X \qquad \ldots 2.51$$

$$\bar{\mu}^R = \mu^R - e\Phi^R \qquad \ldots 2.52$$

$$\bar{\mu}^X = \bar{\mu}^R \qquad \ldots 2.53$$

The changes in electrochemical potential of the electron at each step are

$$Q \text{ to } P: \quad -e\Phi^{vX} - (-e\Phi^{vR}) \qquad \ldots 2.54$$

$$P \text{ to } P': \quad \bar{\mu}^X - (-e\Phi^{vX}) \qquad \ldots 2.55$$

$$P' \text{ to } P'': \quad \bar{\mu}^X - \bar{\mu}^X = 0 \text{ at equilibrium} \qquad \ldots 2.56$$

$$P'' \text{ to } Q'': \quad \bar{\mu}^R - \bar{\mu}^X = 0 \qquad \ldots 2.57$$

$$Q'' \text{ to } Q' \quad \bar{\mu}^R - \bar{\mu}^R = 0 \qquad \ldots 2.58$$

$$Q' \text{ to } Q: \quad -e\Phi^{vR} - \bar{\mu}^X \qquad \ldots 2.59$$

The electrical potential difference

$$V_{RX} \equiv \Phi^{vX} - \Phi^{vR} \qquad \ldots 2.60$$

is called the Volta potential difference between uniform surfaces X and R.

$$_eW^X = -e\Phi^{vX} - \bar{\mu}^X \qquad \ldots 2.61$$

$$_eW^R = -e\Phi^{vR} - \bar{\mu}^R \qquad \ldots 2.62$$

In thermostatic equilibrium, the electrochemical potential for electrons should be the same throughout the two metal systems,

$$\bar{\mu}^R = \bar{\mu}^X$$

$$_eV_{RX} = e\Phi^{vX} - e\Phi^{vR} + (\bar{\mu}^X - \bar{\mu}^R) \qquad \ldots 2.63$$

$$= \left(-e\Phi^{vR} - \bar{\mu}^R\right) - \left(e\Phi^{vX} - (\bar{\mu}^X)\right) = {_eW^R} - {_eW^X} \qquad \ldots 2.64$$

Finally,

$$V_{RX} = \Phi^{vX} - e\Phi^{vR} = W^R - W^X \qquad \ldots 2.65$$

2.13 PRINCIPLE OF MEASURING CPD

When two metals are in contact, and with the assumption $\phi_2 > \phi_1$, electrons of higher electrochemical potential in metal 1 will flow to 2. This process will continue until the electrochemical potentials of the two metals are equal. Thus, metal 2 becomes more negative than metal 1 by an amount equal to the difference in the work function between the two metals. This potential will appear as CPD. Since the contact potentials are a result of equilibration, they are not the e.m.f.'s and thus cannot be measured with the conventional voltmeter (energy required for the measurement can never be withdrawn). The contact potential difference, V_{CPD} between the two metals (Fig. 2.5) in contact is

$$qV_{CPD} = \phi_2 - \phi_1 \qquad \ldots 2.66$$

This electric field induces a quantity of $-\sigma$ on metal 2 and $+\sigma$ on metal 1. Using an external voltage source V_B, σ can be adjusted to zero; under this condition, $V_{CPD} = -V_B$. The condition of thermodynamic equilibrium is of utmost importance in the determination of CPD between two materials.

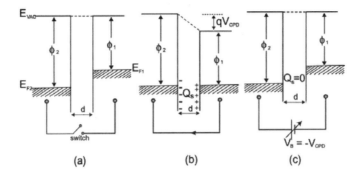

Fig. 2.5 Principle of CPD Measurement of a metal – metal system.

If different conductors $i, j, l, \ldots$ are connected in series to form a close loop with a voltmeter, then the algebraic sum of CPD's in the circuit is zero. So, therefore any type of voltmeter can't measure CPD of a particular junction.

2.13.1 Metal-Semiconductor System

Consider a metal and n-type semiconductor with true work functions $e\Phi_m$ and $e\Phi_s$ respectively; assume $\Phi_s < \Phi_m$. The semiconductor work function is composed of the electron affinity $e\chi$ and the energy difference between the Fermi level (E_F) in the bulk and the conduction band, i.e., qV_s. Upon (external) contact, the E_F's equalize; a negative charge transfer takes place from the semiconductor to the metal because $\Phi_s < \Phi_m$ (Fig. 2.6). This leads to the formation of a positive space charge in the bulk and an upward band bending at the surfaces of the semiconductor.

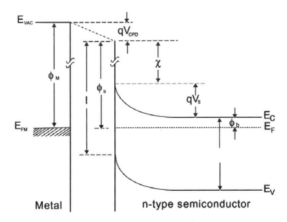

Fig. 2.6 CPD measurement: Metal-semiconductor system.

However, when a large density of surface states are present at the semiconductor surface, then at equilibrium, the charge in the surface states is compensated by those in the space charge layer, the surface states being filled up to E_F. Upon contact with the metal, E_F in the semiconductor falls to an amount equal to qV_{CPD}, if the density of surface states is large enough. The space charge layer and thus the band bending remains unaffected by the contact, *i.e.*, all the electric field lines terminate at the surface leaving the Fermi level in the semiconductor pinned.

For a *p*-type semiconductor, similar arguement holds good.

2.14 NON-UNIFORMITY OF REAL SURFACE AND AVERAGING OF THE KELVIN PROBE

The CPD has been experimentally shown to be highly dependent upon the nature of the surfaces, surface cleanliness, surface roughness and orientation (the aspects related to the anisotropy of the surface are discussed later). Many surfaces consist of a number of areas of different crystal orientations; these surfaces are called "patchy" and these areas are called patches. Surfaces having patches, have non-uniform work function.

Kelvin probe determines the mean work function of a surface and not that biased towards low work function patches as in thermionic emission, photoelectric emission and field emission experiments. The value of work function measured by photoelectric and thermionic emission gives weighed mean that favours the patches of lowest work function.

2.15 ANISOTROPY OF THE WORK FUNCTION

The work function even for a single crystal consists of two parts: (*i*) due to the bulk of the single crystal and (*ii*) due to the 'double layer' at the surface. The 'double layer' is common for metals and the work function due to the double layer depends upon the crystal plane orientations. For example, for tungsten, the faces can be arranged according to decreasing work function as follows: 110, 211, 100 and 111.

The double layer for a metal surface can be atomically smooth or atomically stepped surface (Fig. 1.5). Additional surface double layer (Fig. 1.6) arises out of adsorption. Changes in the charge distribution occur in such a way that a dipole moment *P* can be associated with each adsorbate atom. The adsorbate layer contribute a term $\Delta\varphi$ to the work function

$$\Delta\varphi = 2\pi P N \theta \qquad \ldots 2.67$$

where N is the maximum number of adsorption sites per unit area and $\hat{e}$ is the fraction of occupied sites.

The double layer may also be associated with a polyhedron ("s-polyhedron") with the atom at its center, such that it contains all points nearer to the atom under consideration. If the distribution of the electron density within these polyhedra of the surface atoms is the same as for the inside atoms, then there would be no double layer on the surface. However, this cannot be the case since the total energy is lowered by a redistribution of the electron cloud on the surface. There are two effects: (*i*) partial spread of the charge out of the *s*-polyhedra and (*ii*) a tendency to smooth out the surface of the polyhedra. In consequence of the second effect, the surfaces of equal charge density are more nearly plane than in the original picture. The two effects have opposite influences and since they are comparable in magnitude, it is not possible to predict the sign of the total double layer without numerical computations. Generally, one observes differences between the work functions for different crystallographic directions / planes (Smoluchowski, 1941).

Work functions of stepped metallic surfaces (Fig. 1.6) are modeled to generate work function anisotropy maps. It has been shown by Fall *et al.* (Fall *et al.* 2002) how the work function of any stepped surface can be accurately predicted by interpolating between the work functions. This technique is applied to the work-function anisotropy of tungsten (Fall *et al*, 2002).

Table 2.1

Metal	Work function φ /eV		Metal	Work function φ /eV		
	Photoelectric	C.P.D.		Thermionic	Photoelectric	C.P.D.
Li ...	—	2.32	Nb ...	4.30	—	4.37
Na ..	2.36	2.46	Mo ...	4.33	4.49	4.21
K ...	2.30	2.01	Ta ...	4.33	4.30	4.22
Rb ...	2.05	—	W ...	4.55	4.55	4.55
Cs ...	1.95	1.82	Re ...	4.72	—	—
Be ...	—	3.91	Ti ...	4.10	4.33	4.20
Mg ...	—	3.61	Cr ...	4.60	4.44	—
Ca ...	2.87	—	Mn ...	—	4.08	—
Ba ...	2.52	2.35	Fe ...	—	4.60	4.16
			Co ...	—	4.97	—
Zn ...	3.63	4.11	Ni ...	5.24	5.15	5.25
Cd ...	—	4.22				
			Zr ...	4.00	—	—
Al ...	4.28	4.19	Hf ...	3.65	—	—
Ga ...	4.35	—				
In ...	4.08	—	Ru ...	—	4.71	4.73
			Rh ...	4.72	—	—

Contd.

Metal	Work function φ /eV		Metal	Work function φ /eV		
	Photoelectric	C.P.D.		Thermionic	Photoelectric	C.P.D.
Sn...	4.28	4.43	Pd...	—	—	5.40
Pb...	4.25	3.83	Ir...	4.57	—	—
Cu...	4.65		Pt...	5.36	5.63	—
Ag...	4.26	4.29		4.51		
Au...	5.10	5.28	Th...	—	—	3.71
As...	4.79	—	U...	3.47	3.47	3.63
Sb...	4.56	—	C (dag)...	—	—	4.65–5.0
Bi...	4.34	—	Si...	—	4.95	4.75
			Ge...	—	5.15	4.83

Work function Values from the articles by J. C. Rivière, *Solid State Surface Science* (ed. Mino Green), Vol. 1, 1969 (Marcel Dekker, New York), and by J. Hölzl and F. K. Schulte, *Springer Tracts in Modern Physics*, **85**, 1 (1979).

Chapter 3

Design of the Kelvin Probe and the Methods of Measurement of Contact Potential Difference (CPD)

3.1 INTRODUCTION

The Kelvin probe equipment measures the surface work function and surface photo voltage (SPV). It is being employed in several emerging applications: sensors, self-assembled monolayers and bio interfaces etc. This chapter contains detailed description of the construction and the electronic circuitry of the Kelvin probe instrument. The operational intricacies and the complexities involved in the analyses of the data and appropriate experimental conditions for getting reasonable and realistic measurements are included. The design detailed in here has actually been constructed and tested. It should be mentioned that this design is not unique, but is a working design and a design which can incorporate Photo-emission Yield Spectroscopy (PEYS) and Surface Photo Voltage (SPV) measurements. Also, this is a cost effective design achieving a reasonable accuracy of the order of a few milli electron volts in CPD values.

The design of a Kelvin probe basically depends on three factors: (i) mechanical aspects of the capacitance modulation, detection of Kelvin probe signal using appropriate electronic circuitary, (ii) analysis of the Kelvin signal to extract the contact potential difference (CPD) value, and (iii) selection of proper reference electrode for reactive adsorbates. Depending on the type of application, the design factors are to be optimized. A UHV application generally demands the use of low degassing materials for probe construction and electrical connections. The noise level and choice of the reference electrode plays a key role in getting realistic values of CPD with reasonable sensitivity. For evaluation of reaction kinetics, topographic, spectroscopic and time resolved measurements, fast CPD detection technique is a requirement.

Some of the factors that complicate the interpretation of the data obtained from even a well-made Kelvin probe instrument are: the surface dipoles and adsorbed layers and multi-layer systems that may have ohmic contacts etc. Among the several operational difficulties, the minimization of environmental noise and stray capacitance, and the reliability of a calibration procedure for the Kelvin probe tip, do demand serious attention.

The critical part of the Kelvin probe design is the vibrating capacitor formed between the sample and a periodically modulated reference electrode which generates the Kelvin signal.

Lord Kelvin in 1898 first measured the contact potential difference between copper and zinc electrodes using gold leaf electrometer (Fig. 3.1) (Kelvin, 1898). In the equilibrium state illustrated by Fig. 2.5(*b*), CPD is represented as the local modification of the vacuum level and has the magnitude, $V_{CPD} = \Phi_1 - \Phi_2$. The electric field created by V_{CPD} induces charges of opposite sign on the capacitor electrodes. This induced charge was detected (using a gold leaf electroscope) as a transient current flow in an external circuit by changing the spacing between electrodes momentarily. The individual charging and discharging cycles of operation for each test bias potential is rather tedious, time consuming and can introduce significant errors in the measurement. This experimental technique was improved by Zisman by introducing a periodic modulation technique of the capacitor spacing and a three stage audio amplifier and head phone for signal detection (Zisman, 1932). The periodic modulation is commonly referred to as vibrating capacitor. In this technique, by varying the spacing between the electrodes sinusoidally, an alternating current is induced, and the resulting voltage drop across a sensing resistor was then nulled by introducing variable d.c source in series with the circuit; the nulling voltage corresponds to the CPD. Most of the later versions of the Kelvin probe designs follow the basic concepts introduced by Zisman.

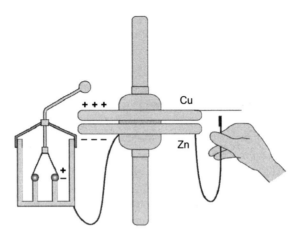

Fig. 3.1 The arrangement used by Lord Kelvin for CPD measurement between copper and Zinc.

Most common reference capacitor plates are: gold, tungsten or stainless steel. The work function of the reference capacitor plate should be stable, should not undergo any adsorption and its surface work function must be measured by an independent experiment, like, Photo Emission Yield spectroscopy (Chapter 4). In studying the corrosion properties, Hg/HgSO$_4$ capillary reference electrode is being used (Kamimura and Stratmann; 2001).

More details of the reference electrode geometries and the procedure adopted to determine their work functions are described in section 3.8

3.2 THE DESIGN ASPECTS OF THE KELVIN PROBE: THE VIBRATING CAPACITOR
3.2.1 Ideal Parallel Plate Capacitor Geometry

For a sinusoidal modulation of the distance between the reference electrode and the sample, $d(t) = d_0 - a\,sin(\omega_0 t)$, the time varying ideal capacitance $C(t)$ can be written as

$$C(t) = \frac{\varepsilon_0 A}{d_0 - a\,sin(\omega_0 t)} \qquad ...3.1$$

A is the effective area of the vibrating capacitor, d_0 is the mean spacing, 'a' is the amplitude of vibration, ω_0 is the angular frequency of vibration and ε_0 is the permittivity of free space.

Equation 3.1 may also be rewritten as

$$C(t) = \frac{C_0}{1 - m\,sin(\omega_0 t)} \qquad ...3.2$$

where $m = a/d_0$, is the modulation index and $C_0 = \varepsilon_0 A/d_0$, is the capacitance in the absence of modulation. The capacitance $C(t)$ contains series of harmonics of ω_0 (because the capacitance is inversely proportional to the spacing between the probe and the sample).

Fourier series expansion of $C(t)$ can be written as

$$C(\omega t) = a_0 + \sum_{n=1}^{\infty}\left[\left(a_n\,cos(n\omega_0 t) + b_n\,sin(n\omega_0 t)\right)\right] \qquad ...3.3$$

The coefficients are obtained using the method of contour integration for $m<1$ and is given by

$$a_0 = \frac{C_0}{\sqrt{1-m^2}} \qquad ...3.4$$

$$a_n = \frac{2C_0}{m} \sin\left((n-1)\frac{\pi}{2}\right) \frac{m^n}{(1+\sqrt{1-m^2})\sqrt{1-m^2}} \quad \text{...3.5}$$

$$b_n = \frac{2C_0}{m} \cos\left((n-1)\frac{\pi}{2}\right) \frac{m^n}{(1+\sqrt{1-m^2})\sqrt{1-m^2}} \quad \text{...3.6}$$

Using equation 3.3, the fundamental component of the capacitance can be written as

$$C(\omega_0 t) = \frac{2mC_0}{\sqrt{1-m^2}\,(1+\sqrt{1-m^2})} \sin\omega_0 t \quad \text{...3.7}$$

This expression is frequently cited in the literature for approximate calculation of current generated by the vibrating capacitor (Germanova, 1987, Rossi, 1992).

In the simple approach of CPD measurement as illustrated in Fig. 3.2, the current generated by the vibrating capacitor can be expressed as,

$$i(t) = \frac{d}{dt}\left\{\left[-V_{CPD} - V_B - V(t)\right]C(t)\right\} \quad \text{...3.8}$$

Measurement procedure for $I(t)$ and its harmonic contents will be discussed later in this chapter.

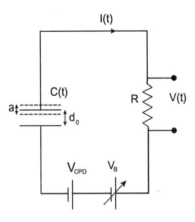

Fig. 3.2 The vibrating capacitor for CPD measurement.

In the practical design of the vibrating capacitor, geometry of the capacitor, capacitive coupling of the vibrating capacitor with its surrounding and non-uniform work function of the capacitor electrodes may cause appreciable deviation from the ideal situation discussed above. All these non-ideal situations may show up in spacing (d_0) dependence of CPD (Ritty et al., 1982).

3.2.2 Effect of Fringe Field and Non-parallelism

While deriving the equation for the current, near ideal and perfectly parallel electrodes are assumed for the Kelvin capacitor. But in practical experimental conditions, there are significant deviations from the near ideality. For example, flat and large area surfaces are preferred for the sample (forming one plate of the capacitor) where as the vibrating reference plate should be very small and sometimes needs to be hemi-spehercial. The size and shape of the reference electrode vary with the specific requirement of the experiment. The sinusoidal mechanical motion of the reference electrode imposes constraints on its mass, size and shape (which govern the mechanical resonance frequency); maintaining near ideal parallelism with the sample, limits the area of the reference electrode. Thus, the two plates of the Kelvin capacitor have inequal areas giving rise to significant fringe field effects (Fig. 3.3). The non-uniformity of the sample surface (like a patchy surface) also gives rise to fringe field effects. The schematic of a perfect plane-parallel capacitor with the electric field lines contained within the plate area and fringing fields is shown in Fig. 3.3(a). The vibrating capacitor in the present design can be modeled as a parallel-plate capacitor with unequal plates as shown in Fig. 3.3(b).

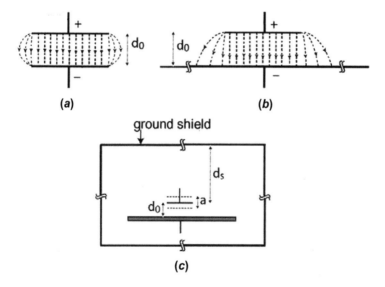

Fig. 3.3. (a) Ideal plane-parallel capacitor with the electric field lines contained within the plate area and fringing fields (b) Practical parallel plate Kelvin probe capacitor geometry, and (c) Screen electrode, sample and vibrating reference electrode in a practical measurement setup (d_s denotes the distance between vibrating electrode and screen).

For a circular disk reference electrode of radius r and negligible thickness separated by a distance d_0 from the sample surface of relatively large area, the static Kelvin probe capacitance (C_0) can be described using the semi-numerical interpolation formula (Chew and Kong, 1980),

$$C_0 = \frac{\varepsilon_0 \pi r^2}{d_0}\left[1+\frac{2}{\pi}\left(\frac{d_0}{r}\right)\ln\left(\frac{1}{2\left(\frac{d_0}{r}\right)}\right)+\frac{3.836}{\pi}\left(\frac{d_0}{r}\right)^2+\frac{6.36}{\pi}\left(\frac{d_0}{r}\right)\right] \quad ...3.9$$

The first term in equation (3.9) is the conventional capacitance for a parallel plate capacitor and the remaining terms represent the contribution of fringing capacitance from top surface and edge of the disk.

The contribution of the top surface and edge to the total capacitance is relatively small for large area reference electrodes. However, precise calculations require a more detailed quantitative analysis for the capacitance (Chew and Kong, 1980; Langton, 1981). The value of C_K, *Kelvin probe mean capacitance* equation 3.10, can be calculated with an error better than one per cent for $\frac{d_0}{r} < 0.5$ using equation 3.9. The fringing field capacitance is of particular importance in the case of reduced probe dimensions and it is larger for $\frac{d_0}{r} \sim 1$. Newton's fringe field capacitance due to edge effect may induce non uniform capacitive interaction. Kelvin probe mean capacitance (C_K) including fringe-field term is given by

$$C_K(d_0) = \varepsilon_0\left(\frac{A}{d_0} + r_p \ln\left(\frac{\sqrt{A}}{d_0}\right)\right) \quad ...3.10$$

where A is the area of the capacitor plate and r_p is the radius of the disc shaped capacitor plate. For spacing much smaller than the typical width of the electrodes, the fringe field contribution can be negligible. A closely spaced guard ring around the vibrating electrode can be employed to minimize the effect of fringe field (Craig and Radeka, 1970; Harris and Fiasson, 1984).

Non-parallelism of the vibrating capacitor generally shows measured value of CPD biased more towards the region having largest capacitive coupling (Baikie, 1988). Lever arm type driving mechanism whose pivot point is at some distance away from the capacitor plates results in time dependent non-parallelism. In normal mode driving mechanism, this effect can be minimized by careful alignment of the capacitor plates. The effect of non-parallelism can also be minimized by operating the probe at small modulation index, m ($a \ll d_0$).

3.2.3 Stray Capacitance Effect

In practice, the Kelvin probe is not an ideal isolated system since it is capacitively coupled to the walls of the chamber, mounting metallic parts including the sample holder. The coupling also is strong between the electrical leads inside the vacuum chamber. The basic questions are: (*i*) under these non-ideal conditions, what is the geometrical equivalent capacitance of the capacitor shown in Fig. 3.3(c) and (*ii*) how to minimize the coupling. The first question is mathematically non-trivial (Langton, 1981). The Kelvin probe signal includes a component due to the stray capacitance. The stray capacitance effect can be minimized by connecting a preamplifier very near to the Kelvin capacitor (Surplice and D'Arcy, 1970, Ritty *et al.*, 1982). A guard ring around the vibrating probe or a variable potential shield has been used to reduce the stray capacitance effect (Craig and Radeka, 1970, Baikie *et al.*, 1991). However, these precautions may give rise to difficulties in sample mounting and manipulation. An analytical compensation technique also is effective to minimize the stray capacitance effect (Hadjadj *et al.*, 1995).

3.2.4 Non-uniformity of the Work Function

The surfaces of metals and semiconductors are known to exhibit non-uniform work function due to patches and other crystallographic imperfections. Such a non-uniformity in the work function of the surface measured by the vibrating capacitor is equivalent to several smaller vibrating capacitors $C_i(t)$ each with different CPD and connected in parallel. The measured CPD in such a model is given by (Ritty *et al.*, 1982),

$$V_{CPD} = -\sum_{i=1}^{N} v_i \left(\frac{A_i}{d_{0i}^2}\right) \bigg/ \sum_{i=1}^{N} \left(\frac{A_i}{d_{0i}^2}\right) \qquad ...3.11$$

This shows the weighing of the different CPDs (v_i's) according to area A_i and spacing d_{0i} of each capacitor element. This weighing property of Kelvin probe is often considered an advantage.

3.3 THE MODULATION METHODS AND NOISE REDUCTION TECHNIQUES: AN OUTLINE OF LITERATURE

Over the years several methods of modulating the Kelvin capacitor and improved signal detection techniques have been developed. Most of these designs are based on the concepts proposed by Zisman (Zisman, in 1932). Different driving mechanisms to modulate the reference capacitor plate have been used: electromagnetic solenoid (Parker and Warren 1962), voice-coil (Baikie *et al.*, 1989), piezoelectric actuator (Besocke, and Berger 1976), and electrostatic actuator (Fain *et al.*, 1976) etc. Peterson used a commercially available piezo-driven tuning fork in his Kelvin probe (liquid-surface) potential sensor (Peterson, 1999). Conventional piezo ceramics also can be used as precision actuators typically operate in the "low voltage" range of 30-150 V. Pendulum type area modulation method was used by Hölzl and Schramenin to monitor the continuous changes in CPD during ion bombardment or deposition (Hölzl, and Schrammen 1974). Baumgärtner and Liess designed a high spatial resolution (40 mm) scanning Kelvin probe using piezoelectric driver with a sensitivity better than 6mV (Baumgärtner and Liess , 1988). Double frequency system described by Blott and Lee gives CPD readings weighed more near to the probe axis (Blott and Lee, 1969). High spatial resolution (50 mm) wire type vibrating reference electrode was used by Butz and Wagner and they achieved a sensitivity of 20 mV in CPD measurement (Butz and Wagner, 1977). UHV compatible piezoelectric driver was introduced by Besocke and Berger (Besocke and Berger;1976) and it is widely adopted by many researchers (Germanova *et al.*, 1987, Lundgren, and Kasemo, 1995, Bellier; *et al.*, 1995; Peterson, 1999). The non-vibrating variant of Kelvin probe has been developed by Qcept Technologies (Qcept Technologies).

Rossi and coworkers used a Kelvin probe with microvolt precision for measuring electric field variations (Rossi *et al.*, 1992). Craig and Radeka used guarded field-effect transistor input stage located within the probe in order to minimize the leads capacitance (Craig and Radeka, 1970). Baikie, et al, have used a voice coil probe with voltage preamplifier to over come the difficulties encountered in phase sensitive detection technique; the out put of the voltage preamplifier is directly coupled to the data-acquisition system (DAS) and additional software techniques are used to reduce the noise (Baikie *et al.*, 1988). A pulsed method of capacitor drive was used by Harris and Fiasion to reduce the driver pick up (Harris and Fiasson, 1984).

The cited literature is only indicative.

3.4 KELVIN PROBE DESIGN

The modulation of the Kelvin capacitor using an electromagnetically driven cantilever mechanism is explained in this book (Suresh *et al.*, 1996). The Kelvin signal detection electronics is based on current preamplifier and phase sensitive detector. The reference electrode and probe assembly are designed to facilitate both CPD and SPV on semiconductor surfaces in vacuum. The reference electrodes used are : stainless steel, gold, tungsten and colloidal graphite coated metal grid. The same reference electrode /probe assembly is adapted for CPD topography in ambient air.

3.4.1 Kelvin Probe Head Mechanical Design

The Kelvin probe reference electrode is modulated by a thin spring steel reed excited by a low power excitation coil (solenoid). The distance between the ferrite core of the solenoid and the reed is adjusted for optimum amplitude of vibration and held in place with the help of 6 mm bolts (Fig.3.4). The use of a solenoid makes the oscillation frequency of the cantilever (holding the reference electrode) twice that of the driving frequency. Thus, detecting the signal at twice the driving signal frequency using lock-in amplifier can effectively minimize/eliminate any coupling of the driving voltage to the genuine Kelvin signal. The vibration amplitude of the probe is sensed using a magnetic induction type vibration probe (a small cylindrical permanent magnet attached to the reed moving inside a solenoid) fixed adjacent to the excitation solenoid. The schematic diagram and photograph of the Kelvin probe are shown in Fig.3.4 and Fig.3.5, respectively. The reference electrode is isolated electrically from the reed and other metallic mounting parts using a ceramic spacer glued using Varian vacuum sealing compound (bakeable to 120°C). The reference electrode is connected to the signal connector with a thin copper wire (1.0 mm diameter). The Kelvin probe assembly has a length of about 10 cm. It is mounted on UHV compatible rotary and linear motion feed through (DNDS 35, Leybold) using the M8 bolt as shown in Fig.3.5. The alignment post provided in the probe head is set initially using a dial gauge (Mitutoyo) in order to maintain the parallelism between the reference electrode and the sample surface.

3.4.2 Vibration Characteristics of the Probe

Frequencies of the fundamental and the enharmonic overtones can be calculated from the formula for cantilever structure clamped at one end (McLachlan, 1951).

$$\omega_n = (k_n l)^2 \left(\frac{a}{\sqrt{3l^2}} \right) \left(\frac{E}{\rho} \right)^{\frac{1}{2}} \qquad \ldots 3.12$$

where $(k_n l)$ are the roots of the equation, $cos(kl) = \dfrac{-1}{cosh(kl)}$, a-thickness, l-length, E-Young's modulus, ρ-density. The ratios of the frequencies of the overtones to that of the fundamental are:

$(k_n/k_1)^2$ 6.27, 17.55, 34.4, 56.8

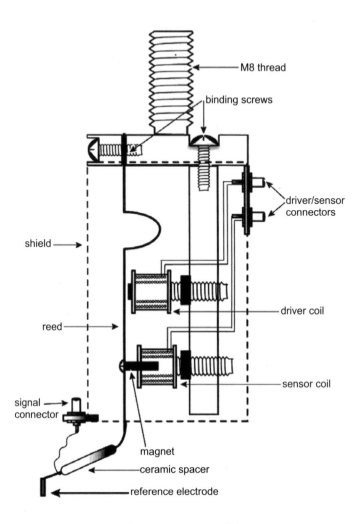

Fig. 3.4. Schematic diagram of the probe.

Fig. 3.6 shows the experimentally observed fundamental and enharmonic overtones of the vibrating reed. . The ratios of the frequencies of the overtones to that of the fundamental are

$$(k_n/k_1)^2 \cong 4.282, \ 5.845, \ \ldots..,$$

The difference in $(k_n/k_1)^2$ values between the theory and the experiment could be due to the shape of the probe tip ("L" shape) at the free end of the reed and additional load due to the vibration sensor magnet. A rapid change in the phase of the oscillation is observed around the resonant frequency

(Fig. 3.6). The operating frequency of the reed is kept away from the resonant frequency in order to achieve phase stability of the output signal. The phase stability is a very important when using lock-in amplifier (LIA) for signal detection.

In order to find the optimum resonance frequency, the frequency is swept in increments 0.5 Hz over a range ± 30 Hz from the estimated resonance frequency while monitoring the amplitude using LIA; a maximum in the amplitude corresponds to the optimum resonance frequency.

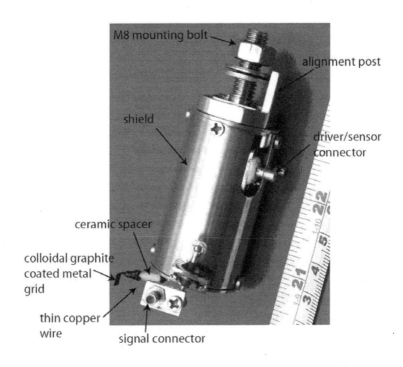

Fig. 3.5. Photograph of the probe head.

While characterizing the probe, d_0 is set with an accuracy of ±2 μm using a high-resolution micrometer head (L. S. Starrett, Co.). The amplitude of vibration of the probe was calibrated in terms of the vibration sensor out put signal. The sensor output has linear dependance on the amplitude of the vibration of the probe (Fig. 3.7). This linearity enables to estimate the amplitude of vibration of the probe. The work function of the colloidal graphite coated metal (tantalum) grid has been estimated using a 3.2-mm diameter stainless steel disk reference electrode employing Surface Photoemission Yield spectroscopy (explained in the next chapter).

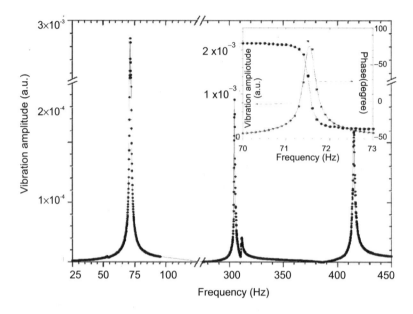

Fig. 3.6. Fundamental and enharmonic overtones of the vibrating reed (experimental results). Insert shows the phase characteristics.

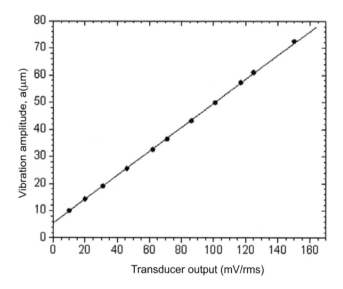

Fig. 3.7. Calibration curve for vibration amplitude of the Kelvin probe.

3.5 DESCRIPTION OF THE MEASURING CIRCUIT

3.5.1 Conversion of the Displacement Current of Vibrating Capacitor into Voltage

The displacement current generated by the vibrating capacitor can be converted into a measurable voltage using a high resistance, R, connected across the vibrating capacitor as illustrated in Fig. 3.2. This arrangement demands the use of a high impedance voltage preamplifier before feeding the signal to the Lock-in amplifier or any other measurement system. The stray capacitance in a practical measurement system such as the capacitance between the probe and its surrounding and coaxial cable capacitance will shunt most of the current $I(t)$ away from R. This problem can be addressed by mounting the voltage preamplifier very close to the vibrating capacitor practically without any electrical leads (Craig and Radeka, 1970; Galbraith and Fischer, 1972). However, this procedure may impose difficulties in applying bias voltage and in sample handling. A possible solution to these difficulties is to use a current to voltage converter (current pre-amplifier) as shown in Fig. 3.8(a) (Bonnet et al., 1977; Rossi, 1992a). A current pre-amplifier provides a "virtual ground" to the Kelvin capacitor, which shorts out any input capacitance, C_i in Fig. 3.8(b). This permits the insertion of a low noise coaxial cable from the probe to the preamplifier, allowing it to be isolated from the vibration. Thus a bais V_B can be applied to the Kelvin capacitor through the non-inverting terminal of the op-amp. The advantages of using current pre-amplifier are deatailed in the literature (Bonnet et al., 1977; Rossi, 1992a).

The current pre-amplifier has been used in the present work. A low input bias current, low input capacitance, high input impedance and low noise are essential for optimum conversion efficiency and sensitivity of the Kelvin signal. The typical features of FET-input electrometer grade op-amp, AD 515 used for constructing the current pre-amplifier and the details are given in Appendix II.

The schematic of the electronics and the data acquisition system for the CPD measurement is shown in Fig.3.9. A Princeton Applied Research model 5210 lock-in amplifier along with the home made current preamplifier forms the heart of the signal detection system. The internal oscillator output of the lock-in amplifier is fed to a power amplifier op-amp (APEX, PA-01) which drives the solenoid. The vibration sensor output is amplified and monitored using an oscilloscope (HP 54615 B). The sensor output is found to be free from harmonics. The FET input op-amp (AD515, Analog Devices) is used for constructing the current pre-amplifier. An amplification factor of about 10^{-9} A/V is provided in the current preamplifier using a low leakage high resistance (glass sealed). The ac current generated by the vibrating capacitor is fed to the inverting input lead of the op-amp which is isolated from the rest of the circuit board using Teflon support. The output of the current preamplifier is further amplified using an instrumentation amplifier IC (AD521, Analog Devices) before feeding the signal to the lock-in amplifier. The vacuum chamber and all other metallic mounting parts share a common ground: the preamplifier ground. Narrow band amplification along with phase sensitive rectification of the output signal eliminates the noise significantly.

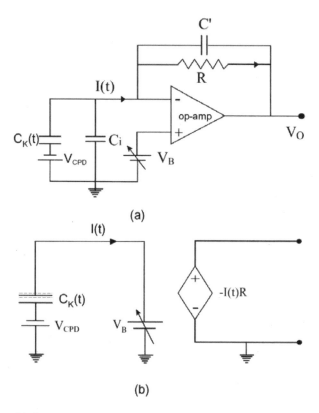

Fig. 3.8 (a) Current pre-amplifier with biasing V_B, and (b) equivalent circuit.

The probe biasing path (non-inverting input terminal of the current pre-amplifier) has the provision for off-null and feed-back mode of CPD measurement. The built-in Digital to Analogue Converter (DAC) of the lock-in amplifier is used for off-null method of CPD measurement. The preamplifier gain and output filter roll-off of lock-in amplifier are adjusted for stable operation in the feed-back mode of CPD detection.

Measurement of CPD is carried out using a Qbasic based computer program which controls the functions of the lock-in amplifier *via* RS232C serial port (better programs can now be written with Matlab, visual basic or Lab View or similar for a USB port of faster data acquisition systems /

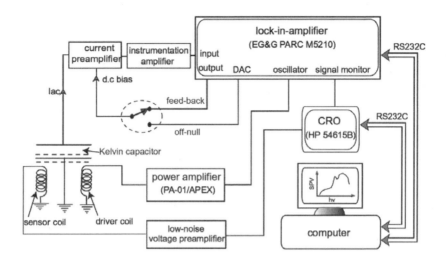

Fig. 3.9 Schematic diagram of the experimental setup.

computers: this is the reason why the explicit program used in the present design is not given). The ADC inputs of the lock-in amplifier are used for monitoring the pressure in the vaccum chamber and temperature of the sample. For the scanning Kelvin probe (for measuring the work function of "patches" and for tography applications), all the mecahical movements (X-Y translation stage of the Kelvin probe, the optical monochromator, shutter control for the optical beams) are controlled using the stepper motors through the parallel-port of the computer. Online data plotting graphic sub-routine is introduced in all the data acquisition programs. Digital storage Oscilloscope, HP54615B is used to record the waveform and analyze its harmonic contents.

3.5.2 The Preamplifier

The signal from the vibrating Kelvin capacitor can be usually amplified using either current or voltage preamplifier. Employing a voltage preamplifier in Fig. 3.10(a), the voltage drop developed across a high resistance, R, is amplified to a level sufficient for sensitive measurement. Any co-axial cable used in between the preamplifier and the Kelvin capacitor makes the stray capacitance, C_i, considerably larger than C_k (Kelvin probe mean capacitance). The stray capacitance C_i discussed earlier does not include the stray capacitance between the probe and the surrounding. Thus, C_i will shunt most of the current I away from R.

On the other hand, the use of current preamplifier presents a "virtual ground" to the Kelvin capacitor and hence significantly eliminates the stray capacitance and facilitates one to apply the bias voltage easily as illustrated in Fig.3.10(b). Also the current pre-amplifier is preferable to voltage preamplifier from the noise point of view (Rossi, 1992). Bonnet *et al.;* and Rossi have explained the advantages of using current preamplifier (Bonnet *et al.*, 1977, Rossi, 1992). In the present design, the current pre-amplifier has been employed.

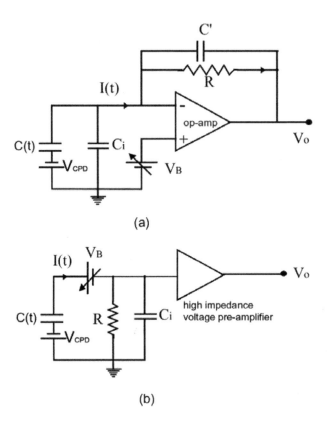

Fig. 3.10 Kelvin probe pre-amplifier configurations (a) high input impedance voltage pre-amplifier, (b) low input impedance current pre-amplifier.

3.5.3 Analysis of the Pre-amplifier Output

In this section, the Fourier components of the current generated by the vibrating capacitor specific to the current preamplifier are examined. Parallel plate capacitor formed between the reference electrode having work function Φ_R and that of the sample having work function Φ_S forms the Kelvin capacitor. The difference in the work functions between the electrodes appears as contact potential V_{CPD}. The basic circuit arrangement is shown in Fig. 3.11. An ac current $i(t)$ is induced in the circuit by varying the spacing between the electrodes periodically and a corresponding voltage drop $v(t)$ is generated in the external circuit.

For the basic circuit arrangement shown in Fig. 3.2, the current $i(t)$ is given by

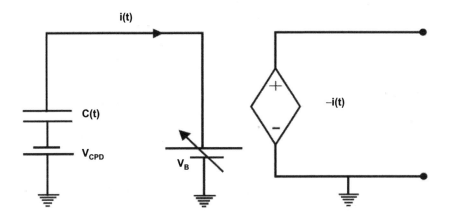

Fig 3.11. Schematic of the experimental setup.

$$i(t) = \frac{d}{dt}\{[-V_{CPD} - V_B - V_i(t)]C(t)\} \qquad ...3.8$$

$V_i(t)$ is the voltage drop across R. For the current preamplifier configuration shown in Fig. 3.11, the negative feed-back converts the high input impedance of op-amp to a virtual short. Under this condition, the contribution of $V_i(t)$ in equation 3.8 can be neglected. The output voltage of current preamplifier can be written using the concept of virtual ground as

$$V_0 = -i(t)R \qquad ...3.13$$

Thus, equation(3.8) becomes,

$$i(t) = (-V_{CPD} - V_B)\frac{d}{dt}\left(\frac{C_0}{1 - m\sin(\omega_0 t)}\right) \qquad ...3.14$$

The current $i(t)$ contains an infinite series of harmonics of ω_0. The discrete Fourier series expansion of $i(t)$ can be written as

$$i(t) = a_0 + \sum_{n=1}^{\infty}\left[a_n \cos(n\omega_0 t) + b_n \sin(n\omega_0 t)\right] \qquad ...3.15$$

The Fourier coefficients (by contour integration) are given by

$$a_0 = 0 \qquad ...3.16$$

$$a_n = \frac{\left(1 + (-1)^{n+1}\right)nm^{n-1}}{\left(1 + \sqrt{1-m^2}\right)^n \sqrt{1-m^2}}\left[m\omega_0 C_0 (V_{CPD} - V_B)\right] \qquad ...3.17$$

$$b_n = \frac{\left(1+(-1)^n\right)nm^{n-1}}{\left(1+\sqrt{1-m^2}\right)^n \sqrt{1-m^2}} \left[m\omega_0 C_0 \left(V_{CPD} - V_B\right) \right] \qquad ...3.18$$

Equation 3.15 is used to calculate numerical values of current for various modulation indices. It can be shown that the ratio of two harmonics is independent of the CPD value and depends on the modulation index (m). This typical property of the Kelvin probe signal is used by Mackel to control d_0 in their scanning Kelvin probe microscope (Mackel et al., 1993).

For relatively large area samples Fig.3.3(c), the semi-numerical interpolation formula for capacitance as described by equation 3.9 can be used for more accurate description of the signal generated by the vibrating capacitor. Using equations 3.9 and 3.8 for a disk shaped reference electrode of radius r, $i(t)$ can be written as,

$$i(t) = \left(-V_{CPD} - V_B\right)\varepsilon_0 m\omega_0 \left[\frac{\pi r^2}{d_0} \frac{\cos(\omega_0 t)}{\left(1 - m\sin(\omega_0 t)\right)^2} \right.$$

$$\left. + 2r \frac{\cos(\omega_0 t)}{\left(1 - m\sin(\omega_0 t)\right)} - 3.836 d_0 \cos(\omega_0 t) \right] \quad ...3.19$$

The first term in the above equation is the contribution of ideal parallel plate capacitor and the remaining terms represent the contribution of fringing capacitance from the top surface and edge of the disk. For relatively large $\frac{d_0}{r}$, the contribution of second and third terms to the total signal is appreciable.

3.5.4 Harmonic Content of the Output Signal

As a first check, the Kelvin probe set up is to be evaluated for the clean Kelvin signal. For this purpose, a *3.2 mm* diameter polished stainless steel disk reference electrode aligned parallel to the surface of a large area ($2cm \times 2cm$) p-type germanium sample forms the vibrating capacitor arrangement. The mean distance, d_0, is set with an accuracy of $\pm 2\,\mu m$. A low-noise triaxial cable is used for coupling the signal to the current preamplifier. The sensitivity of the lock-in amplifier is adjusted for sufficient amplification of the input signal. The signal monitor output of the lock-in amplifier is fed to the digital storage oscilloscope (DSO) for analysis. Signal filters are avoided as they may alter the wave form. The shape of the signal is very sensitive to the modulation index. The signal amplitude is presented in arbitrary units (a.u.) for wave form and harmonic content analysis. Figures 3.12 and 3.13 show the calculated equation 3.15 and experimentally observed wave form of the Kelvin probe signal for various values of modulation index, m, and modulation frequency, $f = 71.6\,Hz$ respectively. Modulation index is varied by changing the mean spacing, d_0 at constant vibration amplitude of nearly *25 μm*.

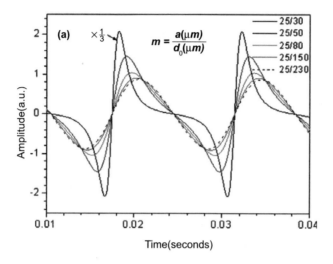

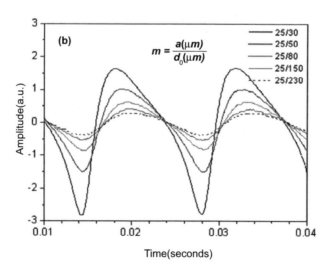

Fig. 3.12 (a) Calculated and (b) experimentally observed Kelvin probe signal for different values of modulation index, m. Frequency of capacitance modulation. $f = 71.6 Hz$

The differences in the shape and symmetry of the calculated and experimental wave forms for larger values of m can be attributed to the non-parallelism and other stray capacitance effects. The frequency dependent gain of current preamplifier (Appendix II) may cause appreciable attenuation for higher frequency components. For small values of the modulation index $(m \approx 0.1)$ the signal is more symmetric and sinusoidal in shape. The Fourier analysis of the signal reveals the relative contribution of harmonics ($f, 2f, 3f$.......) to the total signal.

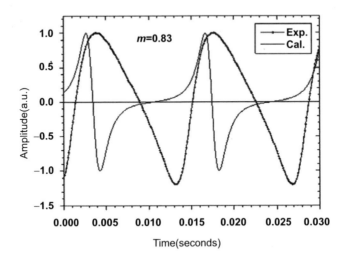

Fig. 3.13 Calculated and experimentally observed Kelvin probe signal at $f = 71.6$ Hz, m=0.83.

Even with all these simplifications, the measured data agrees within 10% to the calculated data for $0.1 < m < 0.5$. The difference in the trend observed between the calculated and experimental curve for f and $2f$ component implies that there is an interference of the microphonic signal with the f component. For high values of m, non-ideal capacitor geometry plays an important role. For large d_0, the vibrating reference electrode behaves like an isolated disk capacitor (low mutual interaction). Using the equations 3.15 to 3.18, it can be shown that,

$$\text{for } a \ll d_0, \quad i(\omega_0 t) \propto \frac{1}{d_0^2} \quad \text{and} \quad i(2\omega_0 t) \propto \frac{1}{d_0^3}, \quad \omega_0 = 2\pi f . \quad ...3.20$$

The form of f and $2f$ component shows that the capacitive interaction of surfaces having different CPD's kept a few centimeters away from the vibrating capacitor will generally have negligible contribution to the genuine signal.

The bias voltage dependence of the harmonic content of the Kelvin probe signal provides an easy way of identifying the interference of electromagnetic pickup from the driving element, microphonetics, stray vibration, power line pickup and other noises with the genuine Kelvin probe signal. Fig. 3.14 shows the harmonic content for $V_B = 0$ and $V_B \approx -V_{CPD}$.

The disappearance of the harmonics when $V_B \approx -V_{CPD}$ clearly indicates the significantly higher signal to noise (S/N) ratio (five orders of magnitude at f). However, it may be noted that this method cannot distinguish the bias dependent capacitive interaction between the Kelvin probe electrodes and other surfaces.

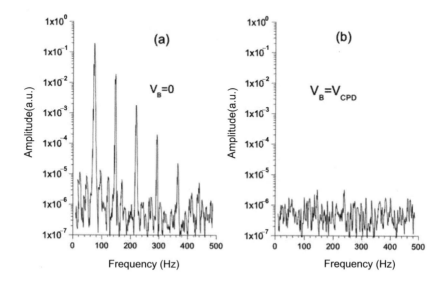

Fig. 3.14 Oscilloscope traces showing the Fourier components of the signal when (a) $V_B = 0$, and (b). $V_B = V_{CPD}$.

No filter is introduced at the lock-in preamplifier stage while analyzing the harmonic content of the output signal. The Harmonic content of the output signal for various values of modulation indices is shown in Fig. 3.15. As may be seen, the harmonic content of the signal depends strongly on the modulation index, m. A rather poor frequency response of the current preamplifier probably is the reason for large attenuation of the higher harmonics observed experimentally. A higher bandwidth is necessary in order to make exact comparison with the calculated Fourier components. Also the non-homogeneous field distribution and deviation from the parallel plate geometry (inclination) may influence the harmonics (Mackel et al., 1993). In the present study, off-null and feed back loop methods have been employed for obtaining CPD under (i) probe grounded and (ii) sample grounded configurations. The shape of the signal is very sensitive to the modulation index (ratio of vibration amplitude and mean spacing). Fig. 3.15 compares the numerically calculated waveform and the one observed experimentally. As shown in Fig. 3.15(b), harmonics of the output signal showed possible interference of coil driving voltage and power line cycle and microphonics roughly five orders of magnitude lower than the fundamental component of the genuine signal.

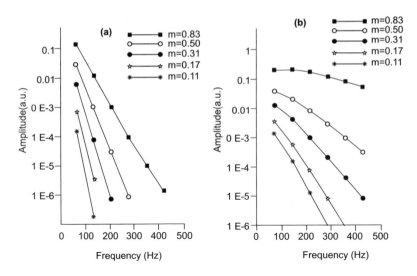

Fig. 3.15 Fourier components for various modulation index, *m*. (*a*) measured signal (*b*) calculated signal *f = 71.6 Hz*. For calculated signal, equation 3.12 is used.

The driver pick up in the present case is negligibly small, mainly because of the low power of the driver coil. By applying a bias voltage nearly equal to the CPD, one can easily identify the genuine signal from the noise (shown in Fig. 3.14). Thus, by analyzing the harmonic content of the preamplifier output signal with and without applying the bias voltage, one can identify the real Kelvin probe signal from the noise. One drawback of this method is that it may not identify the signals generated by the stray capacitance between the vibrating probe and side faces of the sample, the probe mounting metallic parts and vacuum chamber having different CPD's. Since the output current falls as $(1/d_0^2)$, the contribution of other surfaces having different CPD's kept at distances much larger than d_0 will generally have negligible contribution to the output current. This problem has been widely addressed and various suggestions have been put forth: such as the use guard ring around the vibrating probe, grounded vibrating probe configuration and constant potential shield around the Kelvin capacitor to minimize the stray effects (Surplice and D' Arcy, 1970, Ritty *et al.*; 1982, Baikie *et al.*; 1991, Rossi, 1992).

Fig. 3.16 shows the dependence of the current generated in the Kelvin capacitor on the spacing between the capacitor plates and for f and $2f$ modes of vibration; the figure shows both experimental and theoretical results for operating frequencies 71.6 Hz and 411 Hz.

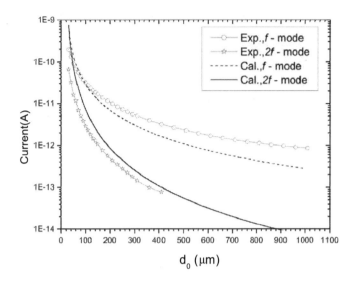

Fig. 3.16(a) Experimentally observed and calculated spacing dependence of f and $2f$ components of vibrating capacitor current (operating frequency 71.6 Hz).

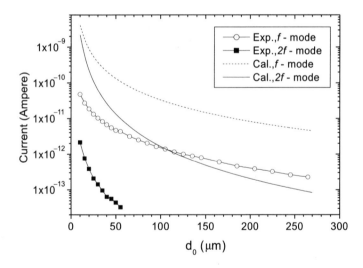

Fig. 3.16(b) Experimentally observed and calculated spacing dependence of f and $2f$ components of vibrating capacitor current (operating frequency 411 Hz).

3.6 METHOD OF CPD MEASUREMENT

3.6.1. Off-Null Method

Using the linear relationship between i_{ac} and V_B (equation 3.19), the null-point is determined from the intersection of the best-fit line with the V_B axis. The backing voltage, V_B is scanned around the null-point using the built in 16-bit DAC of the lock-in amplifier. A good linearity (Fig. 3.17) of the current with bias very close to the balancing point shows the low noise level of the output signal. Off-null method is used every time when a new sample is loaded on to the sample holder and whenever verification with feedback loop CPD value is required.

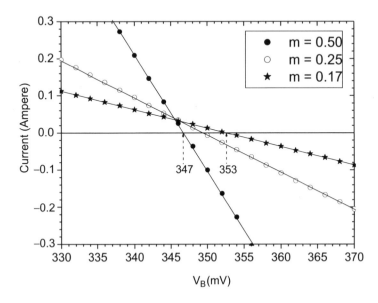

Fig. 3.17. The linearity of the Kelvin probe current as a function of applied bias voltage for various values of modulation index, m.

3.6.2. Feedback Loop System

Continuous changes in the contact potentials during adsorption, temperature variation, optical illumination, topographic application, etc. can be tracked without using feed back loop system. It can be implemented using current preamplifier and a lock-in amplifier with appropriate time constants and slew rates. The output of the current preamplifier after appropriate pre-amplification and filtering and phase adjustment is fed to the non-inverting input of the current preamplifier. The sign of V_B is adjusted by changing the phase setting of the lock-in amplifier in order to null the CPD. Thus, V_{CPD} can be read directly from the output of the lock-in amplifier. Under stable operation of the loop, V_B is given by $V_B = -GV_{CPD}$ where $G<1$ is the D.C. close loop gain.

$$G = \frac{H}{1+H}, \text{ and } H \text{ is defined as the ratio} \left(\frac{-V_B}{V_{CPD}}\right).$$

If $H \gg 1$, $G \sim 1$, then $V_B \approx -V_{CPD}$

It was shown by Rossi that for the fundamental component of the output signal H varies as $\left(1/d_o^2\right)$ and thus V_{CPD} also varies with capacitor spacing d_0; however, insufficient gain in the feed back loop may outweigh other sources of spacing dependence, like, stray capacitance, interaction and spatial non-uniformity of work function etc. (Rossi, 1992). Fig. 3.18 shows the saturation behavior of G with V_B for three samples having different CPD's.

By increasing the lock-in preamplifier gain coupled with the output to a suitable time constant and filter roll-off, one can minimize the spacing dependence of CPD. For Larger spacing (more than 200 μm in the present case) and smaller vibration amplitudes, due to weak signal strength, maintaining the higher gain for stable operation of the feed back loop will be quite impractical.

The response time of the feedback loop system mainly depends on the time constant setting of the lock-in amplifier output signal. Fig. 3.18 (insert) shows the typical CPD response time curve for $1s$ time constant setting. This setting is found to be the optimum for the system in terms of stability and noise. The CPD changes can be tracked within two to three seconds.

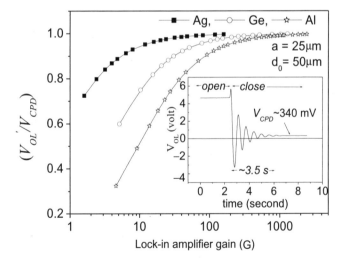

Fig. 3.18 Gain dependence of normalized closed loop bias voltage (V'_{OL}) for samples of different work function values. CPD values measured using off-null technique is used here for normalization (Ag: 0.061 V, Ge: 0.344 V and Al. 0.957 V). Insert shows the typical operation of the feed-back loop.

3.6.3 Effect of Noise

The dc voltage output of the measurement system considered for the off-null and feedback method will have appreciable amount of noise along with the genuine Kelvin probe signal. The bias dependent noise signal arising from the stray capacitance interaction and microphonics in the input signal cable lie exactly at the modulation frequency of the vibrating capacitor and are difficult to eliminate. These are recognized as the dominant sources of systematic error in CPD measurement, and various design procedures for the vibrating capacitor, shielding techniques and signal detection methods have been suggested to minimize the bias dependent noise signal (Surplice and Arcy, 1970; Danyluk, 1972; de Boer et al., 1973; Ritty et al., 1982; Baikie et al., 1991a; Rossi, 1992b, Hadjadj et al., 1995; Dirscherl et al., 2003).

Other sources of noise like (i) broad spectrum of noise sources within the detection electronics (white noise, $1/f$ noise, etc.), (ii) electromagnetic pickup from vibrating capacitor driving element and line voltage and (iii) stray vibrations coupled into the vibrating capacitor, etc., may appear as dc off-set voltage in the CPD measurement. Using off-null method of CPD detection, one can effectively eliminate the influence of all bias voltage independent noise sources without elaborate noise reduction schemes. One can also use filters to reduce the noise; fig. 3.19 shows that the balancing point is found to be independent of the type of filter introduced in the signal detection system.

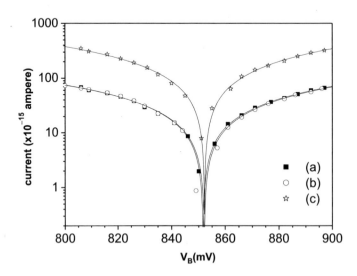

Fig. 3.19 Bias voltage dependence of Kelvin probe signal for different type of filter selection (a) Band pass plus *50/100 Hz* notch filter, (b) all pass filter plus *50/100 Hz* notch filter and (c) all pass filter. (Reference electrode-0.5 mm hemispherical gold tip, sample-thermally evaporated Aluminum on silicon).

However, the sensitivity of the CPD measurement is limited by random fluctuations in the noise whose periods are of the same order of magnitude as the response time of the system. Longer averaging periods may improve the sensitivity. Above all, the actual value of CPD itself can be noisy due to thermal fluctuations on the surfaces and fluctuations in the adsorbates (Palevsky *et al.*, 1947; Rossi, 1992a).

3.7 SPACING DEPENDENCE OF CPD

Since CPD is the difference between the work function of the capacitor surfaces, the nulling bias voltage V_B must be in principle independent of the geometry of the capacitor. In practice, V_B varies with capacitor spacing. Several effects contribute to this spacing dependence. The prominent among them are:

(*i*) non-uniform work function of the capacitor surfaces,

(*ii*) non-parallel capacitor surfaces,

(*iii*) fringe fields,

(*iv*) noise generated by probe driving mechanism and microphonic signal generated by triboelectric and piezoelectric effects in insulators in contact with the signal leads, and

(*v*) insufficient gain in the case of feed back loop detection system.

Widely spaced variable potential shield or a closely spaced guard shield to the vibrating probe was used to eliminate spacing dependence of CPD (Ritty *et al.*, 1982, Harris and Fiasson, 1984, Dirscherl *et al.*, 2003). The setup described here contains only a widely spaced grounded electrostatic shield (vacuum chamber).

In order to avoid problems related to the vibration mechanism, it is preferrable to ground the vibrating reference electrode and have the sample (floating) connected to the input of the preamplifier which can reduce the modulation of the stray capacitance significantly (Surplice and D' Arcy, 1970, Ritty *et al.*, 1982, Baikie *et al.*, 1991). But it is often convenient to have the signal from the vibrating reference electrode when sample surface needs to undergo a change (a heat treatment, deposition of adsorbates and thin films, application of a bias across the sample). In the system described here, a provision has been made for taking the signal from the (floating) vibrating probe.

The spacing dependence of CPD observed experimentally under "*f*" and "*2f*" mode of operations, respectively for sample grounded and probe grounded configurations are shown in Figs. 3.20 and 3.21 respectively. For a small area of the electrode, < 5 micro meter (diameter), the CPD signal in the f mode is shown in Fig. 3.22.

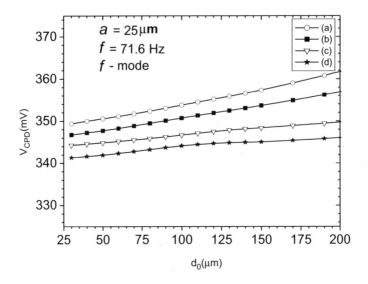

Fig. 3.20 Spacing dependence of V_{CPD} measured at $f = 71.6$ Hz in F-mode. (a) Off-null method with grounded probe. (b) Feed back loop with grounded probe. (c) Off-null method with grounded sample and (d) Feed back loop with grounded sample.

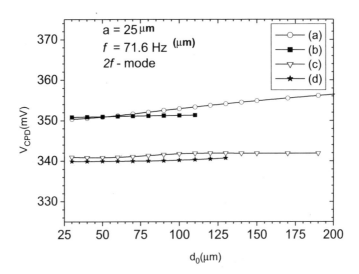

Fig. 3.21 Spacing dependence of V_{CPD} measured at $f = 71.6$ Hz in 2F-mode. (a) Off-null method with grounded probe. (b) Feed back loop with grounded probe. (c) Off-null method with grounded sample, (d) Feed back loop with grounded sample.

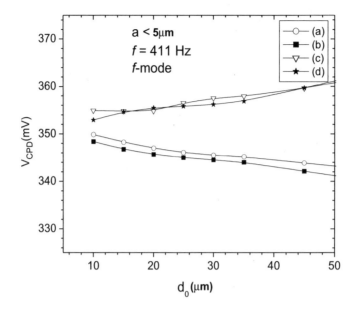

Fig. 3.22 Spacing dependence of V_{CPD} measured at $f = 411\ Hz$ in f-mode for a small area of the electrode ($< 5\mu m$ dia).
(a) Off-null method with grounded probe (b) Feed back loop with grounded probe. (c) Off-null method with grounded sample (d) Feed back loop with grounded sample.

A difference of 10 mV in CPD is observed between the grounded probe and the grounded sample configurations under both "f" and "$2f$" mode of operation. A spurious microphonic signal generated by vibrating signal conductors and insulators that are in phase with the genuine vibrating capacitor signal may be responsible for the systematic difference in the CPD obtained by probe grounded and sample grounded configurations. A negligible spacing dependence was obtained within 0.1 mm of d_0.

3.8 THE REFERENCE ELECTRODE

In order to facilitate absolute work function measurement with the Kelvin probe, a reference electrode of known work function is essential. Also, the reference electrode work function must remain stable with the changes in ambient conditions involved in the course of the experiment. In the literature, many different surfaces have been used as calibration standards, adapted to each case. The surfaces of some materials like barium (Anderson, 1959), highly ordered pyrolytic graphite (HOPG) (Chaney and Pehrsson, 2001; Hansen and Hansen, 2001), gold, tin oxide, stainless steel (Brillson, 1975; Chung *et al.*, 1998) and tungsten (Koenders *et al.*, 1988) show stable work function values in a limited range of ambient condition. The standard electrochemical half-cell such as saturated calomel electrode can be used as reliable work function reference in air (Hansen and Hansen, 2001). For the investigation of band bending in C_{60}/metal interfaces, Hayashi and colleagues adapted the reported work function values of polycrystalline Ca, Mg, Ag, Pb, Cu and Au (Hölzl, 1979) to calibrate their stainless steel reference electrode work function (Hayashi *et al.*, 2002). For passivation studies on GaAs, the reference electrode used is a molybdenum grid (mesh width 250 mm) coated with colloidal graphite having optical transmittance 80% by Suresh and Subrahmanyam (unpublished).

The absolute value of work function of the grid was determined by combining CPD and photoemission yield spectroscopic (PEYS) studies on metallic surfaces such as gold, silver and molybdenum. Investigation of the probe characteristics has been carried out using 3.2 mm diameter stainless steel disk. For topographic application, a hemispherical stainless steel is employed. Baikie used combined measurement of CPD and photoelectron stopping potential measurement (for fixed $h\nu$ above threshold emission) in order to determine the absolute work function of the reference electrode (Baikie *et al.*, 2001).

3.9 VACUUM SYSTEM

The vacuum system for the Kelvin set up should be free from oil to protect the surface from hydrocarbon contamination; a turbo or a cryo vacuum system is preferred. In the present design, a turbomolecular pump (TPU 170, Balzers) with a two stage rotary pump and a cold cathode ionization gauge fixed to the vacuum chamber is used. For all the measurements, the vacuum chamber is evacuated to a pressure of the order of 10^{-9} mbar. The vacuum chamber has three view ports with quartz windows through which the optical beams can be focused on to the sample surface. A heater is located below the sample holder; it facilitates thermally stimulated contact potential difference (TSCPD) measurements also. The sample mounting block has a provision for liquid nitrogen circulation.

The Fig. 3.23 shows the sketch of the experimental chamber. The position of the vibrating electrode is set using a micro manipulator (rotary and linear feed through). The probe is moved away from the sample in order to gain access to photoemission yield experiment integrated with the Kelvin probe. While characterizing the probe, d_0 is set with an accuracy of ±2 μm using a high resolution micrometer head. The amplitude of vibration of the probe is calibrated in terms of vibration sensor output voltage. Cable shielding, vacuum chamber and all other metallic mounting parts are connected to the common earth point: the pre-amplifier input ground. The software is written in Qbasic with online plotting of data.

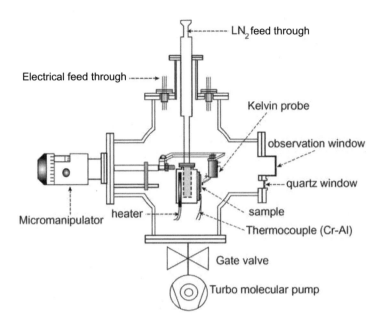

Fig. 3.23 Schematic of the vacuum system.

3.10 DESCRIPTION OF THE SAMPLE HOLDER

The sample holder is designed for the measurement of CPD, SPV and photoemission on the same surface. The schematic diagram and photograph of the sample holder is shown in Figs. 3.4 and 3.5 respectively. The sample holder is made up of nickel-plated copper block having provision for heating using a cylindrical insert type resistive heater and liquid nitrogen circulation for cooling. The sample is isolated electrically from the sample holder with a 300 mm sapphire wafer which is clamped by two electrically insulated stainless steel wires. A chromel-alumel thermocouple mounted on to a reference sample measures the temperature. For intermittent checking of the probe work function, a gold foil mounted on to a stainless steel disk is kept adjacent to the sample.

3.11 SCANNING KELVIN PROBE

Measurement of the surface properties, in particular, the surface work function (or CPD) over a large area gives a comprehensive information about the homogeneity and uniformity of the surface (Forget *et al.*, 2003). The lateral resolution of CPD mainly depends on the probe area and the spacing between the probe and the sample (McMurray and Williams, 2002). The use of a reference electrode of small area allows one to get the CPD topography of the sample surface. A reduction of the probe area results in a weak Kelvin signal. As the signal level decreases, the spacing dependence of CPD plays a very important role in the accuracy of the measurement. Several methods for spacing control have been proposed (Baumgartner, 1992; Palau and Bonnet, 1988). For micro scanning Kelvin probe, it is important to keep the distance constant. In order to achieve nanometer scale resolution in CPD measurement, a modified form of Atomic Force Microscope (AFM) is being developed and is known as Kelvin probe force microscope (KPFM). In KPFM the force detection technique (shift in the resonant frequency of the reed) is being used since the displacement current generated by the KPFM is well below the detection limit of the existing current measurement techniques. For macroscopically defined CPD measurement in the atomic scale, one has to be extremely cautious in the definition of the work function (Okamoto et al., 2003).

Wafer scale CPD, SPV and SPV-Deep Level Transient Spectroscopy (DLTS) measurement using scanning Kelvin probe is employed by Lagel to monitor surface charge, surface potential, and diffusion length of minority carriers to monitor the semiconductor wafer processing (Lagel et al., 1999). A micro-tip scanning Kelvin probe for surface potential measurement on polycrystalline silicon solar cells has ben used by Dirscherl (Dirscherl et al., 2003).

Using the scanning Kelvin probe, local electrical potential gradient in terms of local electrical current density J and chemical potential gradient $\nabla \phi$ has been investigated (Domenicali, 1954). In the absence of any temperature gradient, $\nabla \phi$ can be written as $\nabla \phi = \left(\frac{1}{q} \right) \nabla \mu - \left(\frac{1}{\sigma} \right) J$, where σ is the electrical conductivity. This implies a superposition of electrochemical potential and potential developed due to the resistive property of the material. Employing this concept, the contact potential profile of InGaAs resistors and GaAs MESFETs under bias voltage using KPFM have been investigated (Vatel and Tanimoto, 1995; Matsunami et al., 1999). Contact mode Scanning voltage microscopy (SVM) and Scanning spreading resistance microscopy employs the superposition principle in the investigation of junction properties of semiconductor structures (Ban et al., 2002).

In the present work, the scanning Kelvin probe for surface topographic application having a spatial resolution of 10 μ m is developed using stepper motor driven X-Y translation stage. The scan speed is mainly limited by the feedback loop response time. A hemispherical (0.5 mm diameter) stainless steel probe is used for this purpose. The flatness of the X-Y translation stage is better than 10 mm (tested using a Mittutoyo dial gauge). A negligible spacing dependence of CPD is observed for spacing variations up to few tens of micrometers. Figs. 3.24 and 3.25 give the scanning Kelvin probe CPD topography of aluminum coated p-Si and Al dots coated on p-Si respectively.

The user defines the step (y) and sweep (x) range and the increment to be used for each axis, the positioning of the sample stage and the Kelvin probe for sample sizes up to 100mm × 150mm.

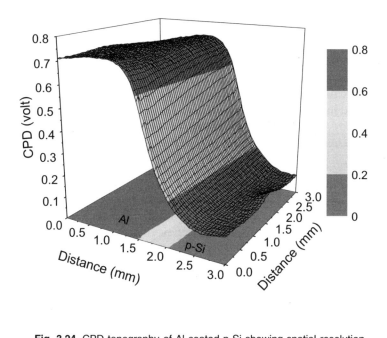

Fig. 3.24. CPD topography of Al coated p-Si showing spatial resolution

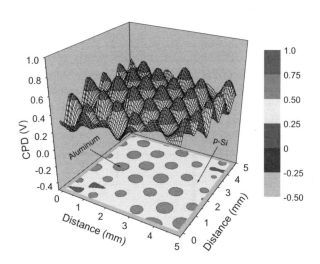

Fig. 3.25. CPD topography of Al Schottky structures on p-Si.

3.11.1 Electrochromic Properties of Tungsten oxide

The Kelvin probe method has been employed to investigate the electrochromic properties of some metal oxide thin films (Subrahmanyam et al., 2006; Subrahmanyam et al., 2007).

In certain large band gap metal oxide thin films, the colour of the films can be switched when a positive ion is intercalated into the lattice. Non-stiochiometric tungsten oxide (WO_{3-x}) is such a metal oxide. When the charge is intercalated into the metal oxide lattice, the Fermi level moves depending upon the charge and these changes in the Fermi level can be measured through CPD using Kelvin probe.

The non-stoichiometric tungsten oxide (WO_{3-x}) thin film is prepared by reactive DC magnetron sputtering technique on indium tin oxide (ITO) coated glass substrate. The color of WO_{3-x} can be modulated from deep blue to transparent by inserting positive ions such as H^+, Li^+, etc., from an electrolyte under the application of external voltage. The change in work function of WO_{3-x} is investigated by intercalating H^+ ions from a 1 Molar aqueous solution of HCl by applying suitable negative voltage ($-2V$ with respect to a platinum counter electrode) to the WO_{3-x} thin film (Fig. 3.26).

Fig. 3.27 illustrates the *CPD* topography of tungsten oxide electro-chromic layer on ITO (scanned over an area of 8×8 *mm²*) depicting the variation in CPD of selectively coloured WO_{3-x} thin film. The CPD value observed for ITO, uncoloured tungsten oxide and colored tungsten oxide thin lms are:- 0.060 mV, +0.075 mV and +0.512 mV, respectively. The change in CPD of the coloured WO_{3-x} shows reduction in work function by approximately 0.5 eV.

Similar experiment carried out on another nearly identical WO_{3-x} film shows almost complete recovery of the film property (work function) after colour/bleach cycle. The region which is subjected to colour/bleach cycle is marked in Fig. 3.28. The blue region of the surface plot in this graph indicates the coloured WO_{3-x}.

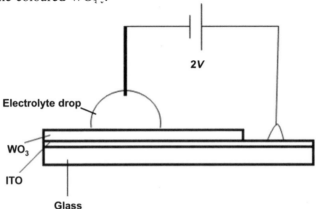

Fig. 3.26 Circuit for colouring an electro-chromic thin film.

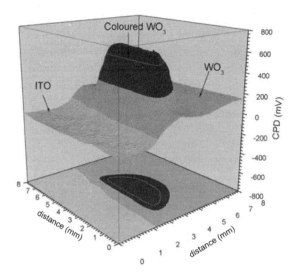

Fig. 3.27 The CPD contour and surface plots of WO_{3-X} / ITO thin film structure. Blue colour region of the colour map surface shows electrochromically coloured WO_{3-x}. Reference electrode is stainless steel.

The mechanism of reduction in work function produced by the H^+ is found to be similar to the inner electrostatic potential modification mechanism (separation of charged species between the conducting solution and the working electrode) in the working electrode of electrochemical cells. The functional relationship between the change in potential, activity, and number of ions can be described in terms of familiar Nernst equation (Janata and Josowicz, 1997). Unlike the metallic electrodes in an electrochemical cell, the WO_{3-x} thin film holds the charges even after removing it from the electrolyte. The inserted charges can be completely removed by shorting the WO_{3-x} electrode with the platinum counter electrode in the presence of electrolyte. It is believed that it is more appropriate to use the term "change in electrode potential" rather than "change in work function" to describe this phenomenon. A quantification of these results may be more useful in evaluating the electro-chromic performance and charge trapping in the tungsten oxide lattice.

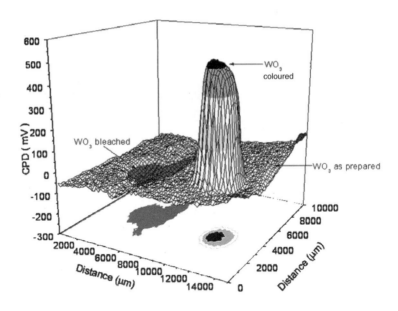

Fig. 3.28 The CPD contour and surface plots of WO_{3-x}/ITO thin film structure showing the electrochromically coloured WO_{3-x} (Blue colour region of the colour map surface) and first coloured and then bleached regions. Reference electrode is stainless steel.

3.11.2 Electrical Potential Gradient in the Presence of Local Electrical Current

Using the scanning Kelvin probe, the superposition of local electrical potential gradient created by an external source with the electrochemical potential of the sample under investigation have been studied (Suresh and Subrahmanyam, unpublished). In the absence of any temperature gradient, the potential gradient, $\nabla \phi$ can be written as (Domenicali, 1954),

$$\nabla \phi = \left(\tfrac{1}{q}\right) \nabla \mu - \left(\tfrac{1}{\sigma}\right) J \qquad \ldots 3.21$$

where σ is the electrical conductivity, J is the current density and $\nabla \mu$ is chemical potential gradient.

One dimensional CPD scan carried out on tin doped indium oxide (ITO) transparent conducting thin film sample with and without external current is shown in Fig. 3.29. The non-uniformity of chemical potential pattern shown as the CPD in curve (b) demonstrates the superposition principle represented by equation 3.21. This technique can be extended to study the phenomenon of electro migration, thermo *e.m.f.*, and semiconductor junction potential (Vatel *et al.*, 1995; Matsunami *et al.*, 1999).

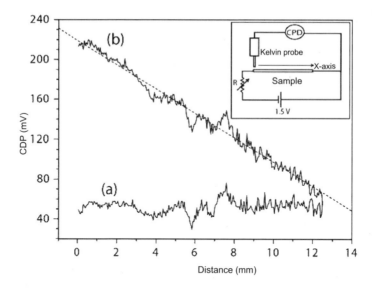

Fig. 3.29. One dimensional line scan of (a) CPD on ITO thin film, (b) CPD gradient induced by the external source of *e.m.f.* Insert shows the measurement schematic.

Chapter 4

Photo-Emission Yield Spectroscopy (PEYS)

4.1 INTRODUCTION

This chapter describes the fundamentals and design of Photo-Emission Yield Spectroscopy (PEYS) integrated into the Kelvin probe set up.

The optical techniques, such as, photo (electro) reflectance (Shen and Dutta, 1995), Reflection anisotropy spectroscopy (RAS) (Weightman *et al.*, 2005), ellipsometry, Raman spectroscopy (Farrow *et al.*, 1987) and Photoluminescence (PL) have immensely enhanced the understanding of the surface electronic structure (McGilp, 1995). A variety of techniques based upon the electrical response and electrical response associated with the optical excitation have been employed to study the surface electronic properties (Brillson, 1982). The spectral dependence of the surface photoconductivity and the surface photovoltage contributed much to the depth of knowledge of the electronic properties of the surface states (Lüth, 1975; Kronik and Shapira, 1999).

Photoemission is the ejection of electrons from the metal or semiconductor surface by the incidence of light of suitable wavelength. These photoemitted electrons carry the information of the surface. The photoelectron spectroscopy (PES) is a powerful tool for studying the electronic properties of bulk semiconductors and their surfaces. The occupied and unoccupied states in two-dimensional surface electronic band structure has been investigated using photoelectron energy distribution measurements: the angle integrated, angle resolved, and yield-type spectroscopy (Feuerbacher *et al.*, 1978). There are several photoemission spectroscopic techniques developed based on the energy of the exciting photon, its polarization, and angle of incidence; the emitted electrons are analyzed for their energy, exit angle and spin polarization. Ultraviolet photoemission spectroscopy (UPS) (Knapp and Lapeyre, 1976; Larsen *et al.*, 1981; Salmon and Rhodin, 1983) and photoemission threshold measurements (Sebenne *et al.*, 1975) are used for identifying the occupied surface states and the structure of valence bands. The unoccupied surface states are probed using inverse photoemission spectroscopy and electron energy-loss spectroscopy (Ludeke and Esaki, 1974).

The availability of the synchrotron radiation has led to the development of new and novel technique: spin polarized photoemission spectroscopy (Feuerbacher *et al.*, 1978). A three-dimensional mapping of the band structure is feasible because of the availability of the continuous spectrum of photon energy. The highly polarized beam of synchrotron source has been used to determine adsorbate sites and molecular orientation on ordered surfaces.

The photoemission yield spectroscopy (PEYS) experimental setup can be integrated with the Kelvin probe equipment. Suresh and Subrahmanyam (unpublished) have investigated the passivation of GaAs surface using PEYS and Kelvin probe; they fabricated the reference electrode made of tantalum metal grid coated with colloidal graphite (Dag coating). The colloidal graphite is well-known for its use in electron physics experiments for reducing the effect of patch fields and to achieve a homogeneous conducting surface having stable work function (Apker *et al.*, 1948; Burns and Yelke, 1969; LeClair *et al.*, 1996; Schedin *et al.*, 1998).

The experimental procedure to estimate the work function of the reference electrode (used as the reference in Kelvin capacitor) and ionization energy of the semiconductor surface is described in this chapter. Emphasis is given mainly to the photoemission process near the threshold region to determine the work function using Fowler's theory (Fowler, 1931). The surface band bending of the semiconductor surface is derived from the measured values of the ionization energy, CPD, and reference electrode work function.

4.2 BASICS OF PHOTOEMISSION FROM SOLIDS

4.2.1 Photoelectric Work Function

The minimum photon energy required for the emission of an electron from its highest occupied energy level in a solid is known as the photoelectric work function, $q\Phi_P$. During this process, the emitted electron acquires the energy from the photon and loses energy $q\Phi_P$ in escaping from the surface. The maximum kinetic energy of the emitted electron, $E_{kin}(max)$ can be related to the incident photon energy, $h\nu$ and $q\Phi_P$ using Einstein's photoemission relation:

$$E_{kin}(\max) = h\nu - q\Phi_P \qquad \ldots 4.1$$

For metals, the highest occupied level is at the Fermi level (E_F) and hence $q\Phi_P$ represents the true work function, $q\Phi$. The semiconductor materials may exhibit $q\Phi_P$ that exceeds its thermionic work function by as much as the energy difference between the Fermi level and the top of the occupied band of energy states. For semiconductors, top of the occupied band is at the valence band maximum (E_{VBM}). Typically, for non-degenerately doped semiconductors, E_{VBM} lies below E_F, in which case $q\Phi_P$ becomes $E_{VAC} - E_{VBM}$ or $\chi + E_g$, the ionization energy, I, and χ is the electron affinity. Occupied surface and bulk states within the band gap results in detectable emission below the photoelectric threshold energy (Gobeli and Allen, 1965; Szuber, 1984). The metal surfaces also show significant contribution in the photoemitted electrons from surface states or resonances (Noguera *et al.*, 1977; Smith *et al.*, 1980).

4.2.2 Photoemission Process

The physical process governing the photoemission is complex. An understanding of this process requires knowledge of the electronic band structure of surface and bulk, various electron energy loss mechanisms and geometrical effects such as reflection and diffraction at the surface. The Fermi's 'Golden-rule' formula for transition probability provides the basis for modeling the photoemission process from solids (Feibelman and Eastman, 1974; Henk *et al.*, 1993; Strasser *et al.*, 2001). The essential features of the photoemission model are: i) the optical excitation of the electrons in the solid, ii) the transport of the electron within the solid, and iii) the escape of the electron through the surface into the vacuum (Berglund and Spicer, 1964). All the three steps are considered together involving initial and final quantum states of the emitted electron by Schattke (Schattke, 1997).

Following the quantum mechanical approach, the photocurrent of the emitted electrons can be expressed in terms of the transition matrix elements of the initial and final states and delta function energy conservation as,

$$i \sim \left(E_{kin.}\right)^{1/2} \sum_{i,f} \left|\left\langle \Psi_f \left| D \right| \Psi_i \right\rangle\right|^2 \delta\left(E_f - E_i - h\nu\right) \quad\quad ...4.2$$

where Ψ_i and Ψ_f are the initial and final state wave functions with energy E_i and E_f, respectively, $E_{kin.}$ is the kinetic energy of the emitted electrons, D the dipole operator and $h\nu$ is the energy of the incident photon, the summation runs over all initial and final states. The direct or indirect transition of electrons in the photoelectric emission process can be related to the optical properties (optical joint density of states) such as reflectivity or absorption coefficient.

In the photoemission process, the optical absorption coefficient together with the emitted electron effective attenuation length (scattering dependent) plays a key role in deciding the relative contribution of the surface and the bulk in the solid (Jablonski and Powell, 2002). The photoemission from surface states, resonances, and surface plasmons are often considered as pure surface photoelectric effect (Endriz and Spicer, 1971). In addition to the single photon photoemission process, two-photon assisted photoemission is also possible with very intense laser radiation through the excitation of surface plasmons (Stuckless and Moskovits, 1989).

4.2.3 Experimental Techniques Based on Photoemission

The photoelectric work function can be derived from retarding potential technique using diode or triode type arrangement of electrodes (emitter, grid, and collector). This method is often difficult to implement in a general purpose experiment chamber because it demands the knowledge of the collector work function, its spatial uniformity and stability (Apker *et al.*, 1948). It is experimentally more convenient to measure the spectral yield of photoelectrons, *Y(hv)*, to determine the photoelectric work function (Sebenne *et al.*, 1975; Szuber, 2000; Yu and Cardona, 2001). The photoelectric work function can be obtained assuming appropriate power law for threshold; requirement of precise knowledge of the photon energy and flux makes this method often more complicated.

4.3 FOWLER'S THEORY OF THRESHOLD PHOTOEMISSION FROM METALS

It is convenient to have a mathematical formulation to accurately predict the threshold energy of the electron emission. In the absence of strong spikes (peaks or shoulders) in the joint density of states near the threshold emission of the photoelectrons, one can conveniently use Fowler's theory of photoemission to find the photoelectric threshold of most of the elemental metals (without invoking the details of transition matrix elements and scattering mechanisms) (Berglund and Spicer, 1964a). Fowler's theory is based on the hypothesis that the number of electrons emitted per quantum of light absorbed is, to a first approximation, proportional to the number of electrons per unit volume of the metal. Based on this hypothesis, the yield near the threshold emission can be written as (Fowler, 1931),

$$Y = AT^2 \left[\frac{\pi^2}{6} + \frac{1}{2}\varepsilon^2 - \left(e^{-\varepsilon} - \frac{e^{-2\varepsilon}}{2^2} + \frac{e^{-3\varepsilon}}{3^2} - \cdots \right) \right] \text{ for } \varepsilon \geq 0 \qquad \ldots 4.3$$

where, $\varepsilon = \dfrac{hv - q\Phi_P}{k_B T}$, and A is a constant.

When $T \to 0$,

$$Y \propto (hv - q\Phi_P)^2 \quad hv \geq q\Phi_P$$
$$= 0 \quad hv < q\Phi_P \qquad ...4.4$$

This approximation is valid over a wide range of temperatures. Fowler's theory accounted well for the near threshold photoemission of most of the metals (bulk and thin films) and also in the description of near-threshold photo-ionization shapes of metallic nanoparticles (Wong and Kresin, 2003).

4.4. BAND STRUCTURE APPROACH TO THE IONIZATION ENERGY OF SEMICONDUCTOR SURFACES

The ionization energy of non-degenerate semiconductor surfaces can be determined from the onset of photoemission from the valence band density of states. Fig. 4.1(a) shows the bulk energy band structure of GaAs illustrating the location of uppermost filled valence band maximum (E_{VBM}) at Γ-point. The location of vacuum level (E_{VAC}) is shown at ~ 5 eV above E_{VBM}. Fig. 4.1(b) shows the density of states in the conduction and valence band. A common method of determining photoelectric threshold, $q\Phi_P$ is to fit a power law of the form,

$$Y \propto (hv - q\Phi_P)^n \qquad ...4.5$$

On the basis of the type of transition involved (direct or indirect) and the scattering processes, Kane (Kane, 1962) has demonstrated that the exponent n ranging from 1 to 5/2 fits the yield curve near threshold (over 0.1 to 0.3 eV range). The high energy photons assume different power laws (Scheer and van Laar, 1963; Gobeli and Allen, 1965). Based on the three step model for the photoemission proposed by Spicer (Berglund and Spicer, 1964) and by Kane (Kane, 1962), Ballantyne (Ballantyne, 1972) derived an equation which gives a good fit to the yield curve over the photon energy ranges of the order of an electron volt.

For non-direct optical excitation in the bulk (Fig. 4.1(a)) and energy loss to quasi-elastic scattering, the yield near threshold can be expressed as (Ballantyne, 1972),

$$Y(hv) = \frac{\alpha(hv)}{\varepsilon_2(hv)} (hv)^{-2} (hv - hv_i)^3 \qquad ...4.6$$

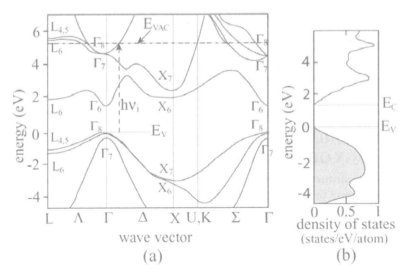

Fig. 4.1. (a) The bulk energy band structure of GaAs illustrating the region of non-direct transition close to E_{VAC} from valence band top at Γ point. The energy zero refers to the valence band maximum. Various energy levels at high symmetry points are represented using symbols (symmetry) with subscript. (b) Density of states in the conduction and valence band (Adapted from Chelikowsky and Cohen, 1976).

where, $h\nu_i$ is the threshold for an indirect transition, $\alpha(h\nu)$ is the absorption coefficient, and $\varepsilon_2(h\nu)$ is the imaginary part of the relative dielectric constant. In the absence of strong structure factor, the term $[\alpha(h\nu)/\varepsilon_2(h\nu)]$ is nearly independent of frequency. Using this assumption, equation (4.6) can be approximated as,

$$Y(h\nu) \sim A(h\nu)^{-2}(h\nu - h\nu_i)^3 \qquad \ldots 4.7$$

$h\nu_i$ corresponds to the photoelectric threshold, $q\Phi_P$, or to the ionization energy, I. Equation (4.7) is adequate to obtain reliable values of ionization of clean and real surfaces of GaAs (Szuber, 1988; Szuber, 2000). For semiconductor surfaces having appreciable density of surface defect states, the threshold may correspond to the position of the uppermost filled electronic state. The threshold may also be altered due to the presence of an electric field (due to surface band-bending) in the region of absorption depth of photons and escape depth of photoelectrons (Kindig, 1967; Fischer and Viljoen, 1971) and the surface photovoltage (Mao et al., 1991).

4.5 MEASUREMENT OF PHOTOELECTRIC THRESHOLD

4.5.1 Experimental Set up

This section describes the experimental set up to measure the photoelectric threshold. The basic Kelvin probe set up, described in Chapter 3, is the starting point. The optical system for photoexcitation, photocurrent measurement electronics, PC based data acquisition and the controls are illustrated in Fig. 4.2. The optical system consists of a *150 W* mercury vapor lamp and a $\frac{1}{4}$-meter grating monochromator (Jarrell-Ash, Model 82-410) and a pair of quartz lenses. The yield measurements are carried out in the wave length range, 2000-4000 Å.

A well-defined image of the monochromator slit is focused ($3 \times 1 mm^2 \, spot$) on to the desired region of the sample surface using the quartz lenses through a quartz window in the vacuum chamber. To minimize photoemission from other parts of the chamber, the beam is oriented so as to strike the surface at near-normal incidence, so that the secularly reflected light can leave the chamber by the same window through which it entered. The detector is either a calibrated spectrometer grade RCA-IP28 photomultiplier tube (PMT) or a diode array. A combination of 0.2 mm slit-widths for both entrance and exit slits give spectral resolution (width at half maximum intensity of 3131 Å spectral line) of about 0.046 eV. The spectral line comparison test for monochromator drive calibration has been made with Ocean optics UV-NIR spectrometer (Model No: HR2000CG).

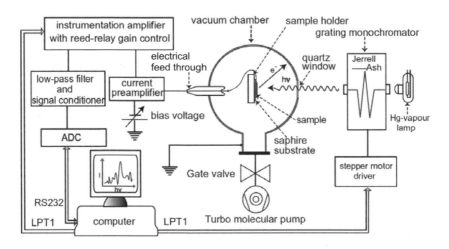

Fig. 4.2.The Schematic of the PEYS setup.

All the metallic mounting arrangements and the walls of the vacuum chamber are grounded to avoid stray photoemission current by the scattered light from parts of the apparatus. The sample is insulated electrically using a 300μm thick sapphire wafer. The photoelectrons emitted from the sample are measured either with a PMT or with the diode type detection system. The emission current is often quite low ($\approx 10^{-12}$ Å), necessitating the use of sensitive and low noise measuring techniques. The emitted photo current is converted into a measurable voltage by a current preamplifier made of low-noise electrometer grade FET input op-amp, AD515 (Analog Devices). The sample is biased at saturation potential through the non-inverting input of the current preamplifier, to ensure that the grounded collector collects all the photoelectrons. An amplification factor 10^{-8} A/V is provided in the current preamplifier, further amplification of the signal is achieved using a variable gain instrumentation amplifier (using computer controlled reed relay setup) and a low-pass filter (~10 Hz, cut-off) cum signal conditioner. The output of the measurement setup is coupled to the computer *via* 16-bit built-in ADC of Lock-in amplifier. The wave length scan is established using a computer controlled stepper motor. The entire measurement is performed using a Qbasic based computer control program.

For evaporated silver film (deposited on silicon oxide: Ag/SiO_2), the typical saturation behaviour of photocurrent with bias voltage is shown in Fig. 4.3. The sample is biased at −12 V for the spectral yield measurements. The electric field strength for the typical bias voltage used is too low to cause appreciable change in the photo threshold due to Schottky effect (Lawrence and Linford, 1930).

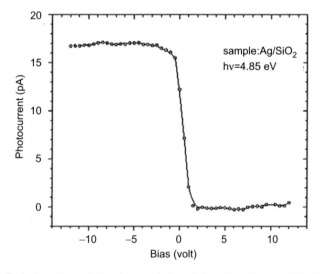

Fig. 4.3. Typical current-voltage characteristics of the photoelectron detection setup for an incident photon energy of 4.85 eV on silver thin film.

4.5.2 Measurement Procedure

Freshly prepared samples are loaded on to the sample holder and the Kelvin probe is set in position for the CPD measurement. During the initial pumpdown of the chamber, due to change in the ambient, CPD is found to drift continuously. After reaching stable values of CPD, spectral yield measurements have been carried out. No observable change in CPD of metallic sample is found while irradiating with UV light from PEYS setup (intensity is too low to produce appreciable photon induced adsorption or desorption of molecules). Pressure in the chamber during the measurement is 4×10^{-7} $mbar$.

Fig. 4.4 shows typical spectral dependence of photocurrent from vacuum evaporated gold (curve a) and silver (curve b) thin films and the photon flux distribution of exciting radiation (curve c) as measured by the IP28 PMT.

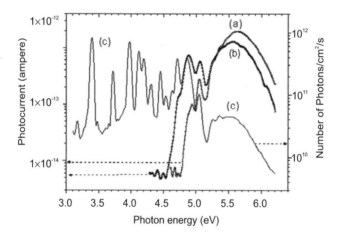

Fig. 4.4. Photoelectric current (left y-axis) as a function of photon energy for vacuum evaporated (a) gold and (b) silver thin films. (c) Spectral dependence of photon flux (right y-axis) measured using RCA-IP28 PMT. Different vertical scales of the curves are indicated by arrows.

The threshold frequency is characterized by a sharp rise in the photocurrent at 4.5 eV and 4.8 eV for thin films of silver and gold, respectively (Fig. 4.4). It may be noted that the threshold frequency depends on the sensitivity of the measurement system. In the present setup, photocurrent values as low as $\sim 5 \times 10^{-14}$ A could be measured without much difficulty. Above the threshold frequency, photocurrent parallels the photon flux distribution. The yield curve is constructed by dividing the photocurrent at each excitation energy with the corresponding flux measured by the PMT. Correction for sample reflectivity (for absorbed photon than incident one) is not necessary in most of the cases. The change in the photo emission yield at the threshold region being two to three orders of magnitude, the associated reflectance change of a few percentages may be neglected.

The heating and cooling cycles and intentional intermittent air exposure can influence the surface conditions, which in turn lead to corresponding changes in CPD and photoelectric threshold. The degree of change observed depends much on the type of the sample. Fig. 4.5 shows the change in the chamber pressure and CPD during heating. The insert shows the heating rate for a set d.c voltage of 110 V for the heater element. The sample holder is provided with a copper block for cooling the sample with liquid nitrogen.

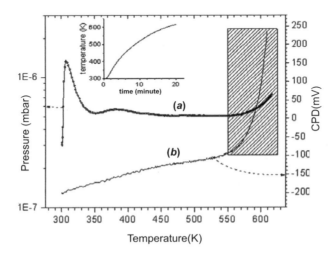

Fig. 4.5. Observed change in (a) chamber pressure and (b) CPD as a function of temperature during heating. Curves (a) and (b) have different vertical scales. Insert shows typical heating rate employed for the heater.

The initial rise in the pressure is due to degassing from the heating element and sample holder. At about 550 K, an increase in the pressure accompanied by a large change in CPD of the sample is observed. The degassing from the inner walls of the chamber and other supporting elements may contribute much to this pressure rise (primarily the water vapour). This typical behavior is observed consistently because of the exposure of the chamber to the atmosphere each time while loading new sample (providing a load lock improves the quality of measurements). The sample heating is limited to within 550 ± 5K in most of the cases. The temperature of the sample is measured using a thermocouple fixed on to a GaAs wafer placed adjacent to the sample under test.

4.5.3 Types of Samples

The samples used to demonstrate the PEYS measurements are:
  (i) Thermally evaporated thin films of gold and silver (99.99% pure, M/s Johnson and Matthey) on thermally oxidized and heavily doped p-

Si substrates. The thickness of the gold and silver film is 2000 Å. The base pressure of the evaporation chamber is 2×10^{-6} mbar.

(*ii*) Gold foil (0.1 mm thick, 99.99% pure, M/s Johnson and Matthey) etched in aqua regia (HCl: HNO_3, 3:1).

(*iii*) Flash evaporated Molybdenum thin film (99.99% pure thermal evaporation source material, M/s Balzers).

(*iv*) n-type Si-doped GaAs (100) wafer with carrier concentration of $\sim 1.7 \, 10^{17} / cm^3$ (M/s Scientific Instruments and Materials Inc, USA)

Cleaning of GaAa : The wafer is subjected to degreasing in trichloroethylene, and acetone in an ultrasonic bath followed by thorough deionized water rinse prior to native oxide etching in dilute HCl. Ohmic contact is made at one corner of the front surface by alloying thermally evaporated indium dots at 350°C in nitrogen atmosphere for one minute. The sample is further subjected to etching in $H_2SO_4 : H_2O_2 : H_2O$: 7:1:1 for minute to remove residual contamination and polishing damage.

4.5.4 Preparation of Reference Electrode in PEYS Experiments

The surface work function of the reference capacitor electrode (for example: gold, molybdenum) in Kelvin probe set up is calibrated by PEYS measurements using colloidal graphite. An aqueous solution of colloidal graphite is deposited by dip coating on to a chemically cleaned square shaped $3.5 \times 3.5 \, mm^2$ tantalum metal grid attached with the ceramic spacer. The graphite coating is initially dried using nitrogen jet followed by hot air. After many cycles of evacuation and exposure to atmosphere over a period of nearly a month, the CPD and PEYS measurements showed stable, consistent and reproducible values.

4.6 STUDIES ON METALLIC SURFACES

The CPD and PEYS measurements are made on thin films of gold, silver and polycrystalline foils of gold and molybdenum to estimate the absolute work function of these materials. Feed-back loop method of CPD detection is employed for monitoring the stability of CPD. CPD is measured using off-null technique (after attaining reasonable stability) to achieve better accuracy. The surfaces of the samples investigated are clean and homogeneous; thus, the photothreshold, $q\Phi_P$ of these metals corresponds to the work function value (Lea and Mee, 1968). The reference electrode work function, $q\Phi_R$ is derived using the relation,

$$q\Phi_R = qV_{CPD} + q\Phi_P \qquad \ldots 4.8$$

Both the CPD and photoelectric work function measurements are made at more than five different conditions for each sample. Measurements repeated at different conditions showed fairly stable values of $q\Phi_R$. From these measurements, the mean work function value of $q\Phi_R$ is estimated using linear regression fitting method.

4.6.1 Gold Thin Film

Gold, being most electro-negative, is less prone to chemisorption of atoms or molecules. However, physisorption related surface dipole layer may cause appreciable shift in the surface work function. Fig. 4.6(a) shows the change in CPD observed from the start of chamber evacuation. During the initial stage, CPD decreases rapidly with time; then slowly increases and attains a saturation approximately after four hours. After attaining stable values of CPD, PEYS measurements are performed. Pressure in the chamber during the measurement is 4×10^{-7} $mbar$.

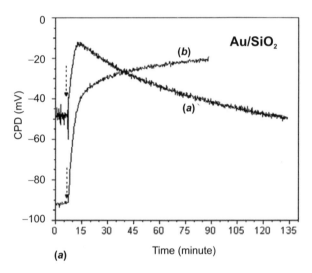

(a)

Fig. 4.6. Change in CPD induced by ambient as a function of time (a) evacuation, and (b) air exposure, Arrow indicates the starting of both the evacuation, and air exposure in the time scale.

Fig. 4.7 shows the yield versus photon energy plot obtained for various conditions, generated by annealing treatment in vacuum. The yield curves are quite similar in their shapes except the one recorded at 110 K. Transitions from the *s*- and *p*-like valence band states just below the Fermi level, contribute to the yield at near threshold (Krolikowski and Spicer, 1970). Above the threshold, a direct transition from the *d* bands dominates the yield curve. The depth of the uppermost *d* band is located at 2.5 eV below the Fermi level (Smith, 1971; Smith *et al.*, 1974). Adsorption related surface modifications expected at various conditions may cause small changes in the electronic structure of the valence band (Krolikowski and Spicer, 1970).

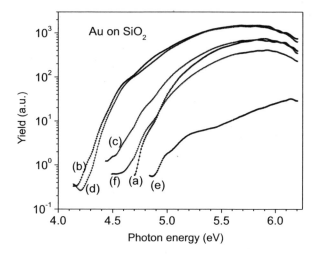

Fig. 4.7. The photoelectric yield in arbitrary units as a function of photon energy. Curves (a) to (f) show data obtained from the same surface of gold thin film (a) after, 4 hours of evacuation, (b) after heated to 250°C and spectra recorded immediately after reaching RT, (c) 24 hours after heating, (d) heated to 250°C second time and spectra recorded immediately after reaching RT, (e) at 110K, (f) after reaching RT.

Fig. 4.8 shows the plot of square root of the yield as a function of photon energy. As per Fowler's analysis of threshold emission from metals, a straight line extrapolation of the linear region of this plot to the abscissa (zero yield) gives the photothreshold, $q\Phi_p$. Low yield tail is excluded from the fit. The values of $q\Phi_p$ obtained for curves (a) to (f) are given in the Fig. 4.8. The standard deviation in $q\Phi_p$ from the extrapolated straight line is about $\pm 0.0015\ eV$ for all the curves. The group of curves (b) and (d) and curves (a), (c) and (f) in Fig. 4.8 show that two successive heat treatments reproduced the same trend in the $q\Phi_p$ change.

The value of $q\Phi_p$ obtained for the air exposed gold thin film at various conditions (curves (a) to (f)) differ as much as 0.3 to 0.7 eV from the clean polycrystalline bulk gold work function of 5.1 eV (Michaelson, 1977). Krolikowski and Spicer reported gold work function as 4.5 eV under poor vacuum conditions (Krolikowski and Spicer, 1970). Wells and Tomlinson reported a work function decrease of 1.0 eV upon exposure of gold to water vapour (Wells and Tomlinson, 1972). Hansen and Johnson have measured the Au work function in air and nitrogen as 4.7 eV using Kelvin probe measurements (Hansen and Johnson, 1994).

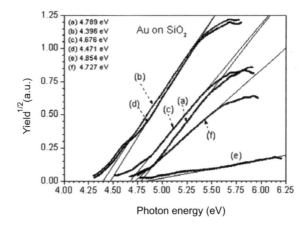

Fig. 4.8. Square root of the yield as a function of photon energy. The straight line in each curve represents the least square fit to the linear region. Curves (a) to (f) show data obtained from the same surface of Au on SiO$_2$ (a) after, nearly 4 hours of evacuation, (b) after heating to 250°C and spectra recorded immediately after reaching RT, (c) 24 hours after heating, (d) heated to 250°C second time and spectra recorded immediately after reaching RT, (e) at 110K, (f) after reaching RT.

The work function modification of gold thin film as a result of the adsorption of gas molecules is rather evident from the range of $q\Phi_p$ values obtained. The large negative shift in $q\Phi_p$ from the clean gold work function of 5.1 eV can be ascribed to the formation of surface dipole layer which in turn lowers the local vacuum level near the surface. The values of qV_{CPD} and $q\Phi_p$ obtained from the above measurements are summarized in Fig. 4.9.

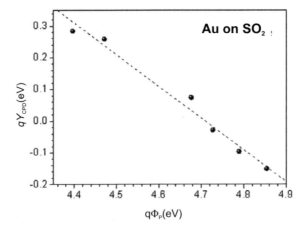

Fig. 4.9. $q\Phi_p$ (determined from the curves (a) to (f) shown in Fig. 4.8) against qV_{CPD} (measured using Kelvin probe) for Au on SiO$_2$. Straight line fit of fixed slope -1 is shown as dotted line.

Fairly stable work function of reference electrode, $q\Phi_R$ is evident from the linearity of the plot. The mean value of $q\Phi_R$ estimated from the linear fit with a fixed slope of -1 is, mean value of $q\Phi_R$ for gold $4.706 \pm 0.026\,eV$

4.6.2 Gold Foil

Gold foil etched in aqua regia is loaded on to the sample holder in the measurement chamber. The experimental approach similar to that used for gold thin film is employed. A change in CPD during initial stages of evacuation shows trend similar to that of gold thin film, marginal difference in magnitude is observed. Fig. 4.10 shows the yield versus photon energy plot obtained for various conditions. As in the case of gold thin film, measurements were

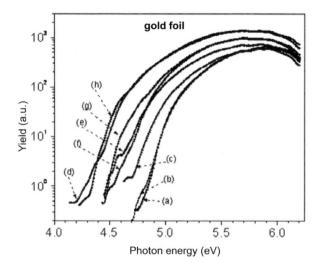

Fig. 4.10. Photoelectric yield in arbitrary units as a function of photon energy. Curves (a) to (h) show data obtained from the same surface of gold foil after, (a) ~ 6 hours of evacuation, (b) ~ 24 hours of evacuation, (c) heated to 250°C and spectra recorded after a time lapse of ~ 24 hours, (d) heated to 250°C second time and spectra recorded immediately after reaching RT, (e) 24 hours after second heating, (f) after isolating the pump from chamber, (g) heated to 250°C third time and spectra recorded immediately after reaching RT, and (h) heated to 250°C third time and spectra recorded immediately after reaching RT.

performed after attaining stable values of CPD. These curves are similar to those observed on the gold thin film. Curves (a) and (b) recorded approximately after 6 and 24 hours; these yield curves are almost identical to those recorded earlier. Thus the stability of the gold surface in high vacuum has been ascertained.

The plot of the square root of the yield as a function of photon energy is given in Fig. 4.11. The values of $q\Phi_p$ obtained for curves (a) to (h) are given in the insert of Fig. 4.11. Curves (d) and (h) recorded after heating the sample to 250°C showed the lowest value of work function. The value of $q\Phi_p$ obtained at various conditions (curves (a) to (f)) differ as much as 0.25 to 0.7 eV from the nominal clean polycrystalline bulk gold work function of 5.1 eV. Over all the behaviour of gold foil is similar to that of the gold thin film.

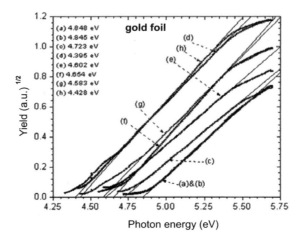

Fig. 4.11. Square root of the yield as a function of photon energy. The straight line in each curve represents the least square fit to the linear region. Curves (a) to (h) show data obtained from the same surface of gold foil after, (a) ~ 6 hours of evacuation, (b) ~ 24 hours of evacuation, (c) heated to 250°C and spectra recorded after a time lapse of ~ 24 hours, (d) heated to 250°C second time and spectra recorded immediately after reaching RT, (e) 24 hours after second heating, (f) after isolating the pump from chamber, (g) heated to 250°C third time and spectra recorded immediately after reaching RT, and (h) heated to 250°C third time and spectra recorded immediately after reaching RT.

The values of qV_{CPD} and $q\Phi_p$ obtained from the above measurements are summarized in Fig. 4.12. It is seen that the linear dependence of qV_{CPD}

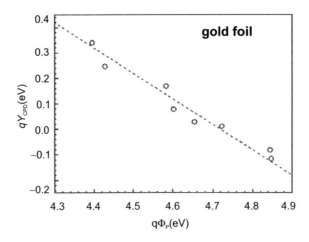

Fig. 4.12. $q\Phi_p$ (determined from the curves (a) to (h) shown in Figure 4.11) versus qV_{CPD} (measured using Kelvin probe) for gold foil. Straight line fit of fixed slope −1 is shown as dotted line.

on $q\Phi_p$ is as good as the one shown for gold thin film in Fig. 4.9. The mean value of $q\Phi_R$ estimated from the linear fit with a fixed slope of -1 is,

Mean value of $q\Phi_R$ for gold foil $= 4.719 \pm 0.012 \ eV$.

4.6.3 Silver Thin Film

The vacuum evaporated silver thin film is loaded on to the sample holder in the measurement chamber. Curve (a) in Fig. 4.13 shows the change in CPD observed from the start of chamber evacuation. During the initial stage, CPD increased slowly from -215 mV with time and then stabilized to a final reading of ~ 255 mV. As in the case of experiments with gold, surface conditions of silver thin film are modified by simple heat treatment. PEYS measurements have been carried out after attaining stable values of CPD. The silver thin film requires settling period of about an hour after the heat treatment. The drift rate is found to be rapid when compared to that of gold. The pressure in the chamber during the measurement is $\sim 4 \times 10^{-7} \ mbar$ all through. The high sensitivity of the silver thin film to the ambient is evident from the observed change in CPD while venting the chamber to ambient air as shown in curve (*b*) in Fig. 4.13.

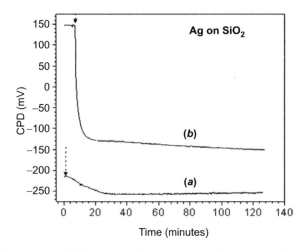

Fig. 4.13. Change in the CPD induced by the change in ambient vs time. (a) evacuation, and (b) air exposure. The arrow indicates the starting of both the evacuation, and air exposure in the time scale.

Fig. 4.14 shows the yield versus photon energy plot obtained for various conditions. The near threshold emission originates from s and p like valence band density of electron states which extends up to 4.0 eV below the Fermi level. The well known d-band is located 4.0 eV from the Fermi level (Smith *et al.*, 1974). In the present case, d-band contribution can be observed for photon energies greater than 8.0 eV (vacuum level dependent). The surface effects play an important role in fixing the vacuum level *via* the formation of surface dipoles. A simple heat treatment produced shift in vacuum level as high as 0.5 eV.

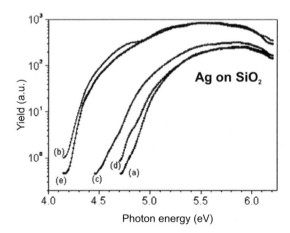

Fig. 4.14. The photoelectric yield in arbitrary units as a function of photon energy. Curves (a) to (e) show data obtained from the same surface of silver thin film after, (a) ~ 2 hours of evacuation, (b) heated to 250°C and spectra recorded immediately after reaching RT, (c) ~ 24 hours after second heating, (d) ~ 48 hours after second heating, (e) heated to 250°C second time and spectra recorded immediately after reaching RT.

Fig. 4.15 shows the square root of yield versus photon energy plot derived from the yield curves. Low yield tail is found to be small compared to that of gold. The values of $q\Phi_p$ obtained using Fowler's analysis of

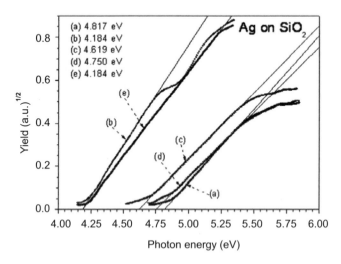

Fig. 4.15. Square root of yield versus photon energy plot. The straight line in each curve represents the least square fit to the linear region. Curves (a) to (e) show data obtained from the same surface of silver thin film after, (a) ~ 2 hours of evacuation, (b) heated to 250°C and spectra recorded immediately after reaching RT, (c) ~ 24 hours after second heating, (d) ~ 48 hours after second heating, (e) heated to 250°C second time and spectra recorded immediately after reaching RT.

threshold emission from metals for curves (a) to (e) are given in the figure. Air exposed fresh silver film displayed the highest work function (4.817 eV, curve (a)). Clean silver has a photoelectric threshold of 4.20 eV (Michaelson, 1977). The observed deviation of 0.6 eV can be attributed to strong dipole layer of oxide layer that is believed to be formed on silver surface in ambient air. Simple heat treatment significantly reduced the work function to 4.18 eV; heating for 24 hours has increased it to 4.619 eV. Farnsworth and Winch, and Anderson reported $q\Phi_p$ values in the range 4.47 eV to 4.79 eV on various poly and single crystalline surfaces of Ag (Farnsworth and Winch, 1940; Anderson, 1941). For (111) oriented planes of vacuum evaporated Ag on quartz plate, Uda and coworkers observed the work function value of 4.35 eV (Uda et al., 1998).

The values of qV_{CPD} and $q\Phi_p$ obtained for silver thin film from the above measurements are summarized in Fig. 4.16. The mean value of $q\Phi_R$ estimated from the linear fit with a fixed slope of -1 is,

Mean value of $q\Phi_R$ *for silver thin film* $= 4.729 \pm 0.015 \ eV$

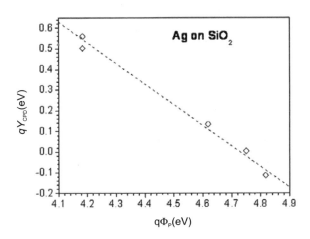

Fig. 4.16. $q\Phi_p$ (determined from the curves (a) to (e) shown in Fig. (4.15)) plotted against qV_{CPD} (measured using Kelvin probe) for silver thin film. Straight line fit of fixed slope -1 is shown as dotted line.

4.6.4 Molybdenum Thin Films

Molybdenum foil cleaned by heating in high vacuum is loaded on to the sample holder in the measurement chamber. The experimental approach is similar to the one employed for gold and silver. The curve (a) in Fig. 4.17 shows the change in CPD observed from the start of chamber evacuation. During the initial stage, CPD decreased rapidly from -180 mV with time and then slowly stabilized to a final value after prolonged evacuation (more than two hours). As in the case of experiments with gold and silver, surface conditions are modified by simple heat treatment. PEYS measurements have been carried out after attaining stable values of CPD. Pressure in the chamber during the measurement is $\sim 4 \times 10^{-7}$ mbar.

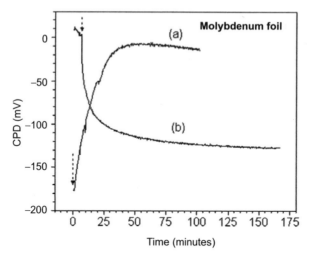

Fig. 4.17. Change in the CPD values induced by the change in ambient against time for Molybdenum foil. (a) evacuation, and (b) air exposure. Arrow indicates the starting of both the evacuation, and air exposure in the time scale.

The yield versus photon energy plot obtained for various conditions are shown in Fig. 4.18. The yield is found to be approximately one order lower compared to that of gold and silver. Also, it may be noted that the change in

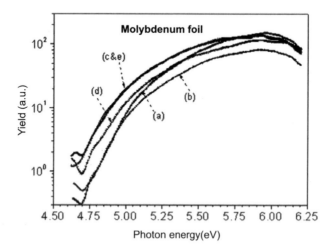

Fig. 4.18. Photoelectric yield in arbitrary units as a function of photon energy. Curves (a) to (e) show data obtained from the same surface of Molybdenum foil after, (a) ~ 4 hours of evacuation, (b) prolonged evacuation (more than 24 hours), (c) heated to 250°C and spectra recorded immediately after reaching RT, (d) ~ 24 hours after heating, (e) heated to 250°C second time and spectra recorded immediately after reaching RT.

the magnitude of yield with treatments is lower compared to that of gold and silver. Direct optical transition of electrons from the large density of states (DoS) associated with $(4d)^5$ and $(5s)^1$ states near E_F contribute much to the yield near threshold (Petroff and Viswanathan, 1971). The peak in the DoS for Molybdenum lie very close to E_F (Bacalis *et al.*, 1985).

Fig. 4.19 shows the square root of yield versus photon energy plot derived from the yield curves. The values of $q\Phi_P$ obtained using Fowler's analysis of threshold emission from metals for curves (a) to (f) are given in the figure. Air exposed fresh molybdenum foil after approximately 4 hours of evacuation displayed the highest work function (4.785 eV, curve (a)). A simple heat treatment resulted in $q\Phi_P$ reduction of ~ 0.15 eV. The overall change in $q\Phi_P$ observed is much smaller than that observed with gold and silver. Molybdenum seems to be a better candidate for reference electrode in vacuum environment when compared to gold. Work function value of clean polycrystalline molybdenum reported in the literature (using various measurement techniques and vacuum conditions) varies from 4.22 eV to 4.49 eV (Eastman, 1970; Michaelson, 1977; also *see* Table 2.1). Among the various gases in the ambient, molybdenum is more sensitive to carbon monoxide (Clewley *et al.*, 1971). In spite of its relatively good stability in vacuum, it is rather sensitive to air ambient as shown in Fig. 4.17.

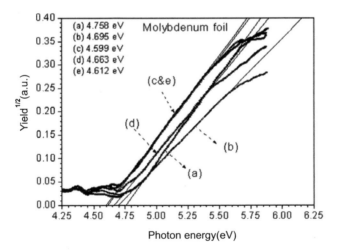

Fig. 4.19. The square root of yield versus photon energy plot. The straight line in each curve represents the least square fit to the linear region. Curves (a) to (e) show data obtained from the same surface of molybdenum foil after, (a) ~ 4 hours of evacuation, (b) prolonged evacuation (more than 24 hours), (c) heated to 250°C and spectra recorded immediately after reaching RT, (d) ~ 24 hours after heating, (e) heated to 250°C second time and spectra recorded immediately after reaching RT.

The values of qV_{CPD} and $q\Phi_P$ obtained for molybdenum foil from the above measurements are summarized in Fig. 4.20. The mean value of $q\Phi_R$ estimated from the linear fit with a fixed slope of -1 is,

Mean value of $q\Phi_R$ *for Molybdenum foil* $= 4.694 \pm 0.007 \ eV$

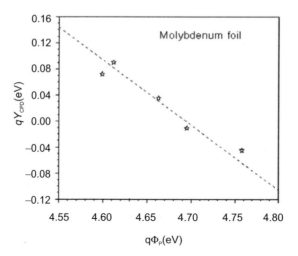

Fig. 4.20. $q\Phi_P$ (determined from the curves (a) to (e) shown in Fig. 3.19) plotted against qV_{CPD} (measured by Kelvin probe) for molybdenum foil. Straight line fit of fixed slope –1 is shown as dotted line.

In the above measurements, the possibility of small changes in the probe work function cannot be ruled out. The observed work function values are comparable to those reported by other workers (Michaelson, 1977, Table 1.3). With more controlled experiments, the differences can be attributed explicitly to the bulk and surface properties of the samples.

4.6.5 Reference Electrode Work Function: Graphite

Graphite is a good reference electrode for measurement of absolute work function of many metal and semiconductor surfaces with Kelvin probe.

Graphite is a layered semiconductor with zero activation energy ($E_g = 0 eV$).

The bulk electronic structure is well approximated by the electronic structure of one monolayer (Bianconi *et al.*, 1977). Since the density of filled valence band states near Fermi level is relatively small than the common elemental metals, the yield near threshold for graphite is found to be lower than that for metals (Taft and Apker, 1955). The fused ring structure of a graphite layer is known to be very stable and inert to chemisorption of common gas molecules of air ambient (Lander and Morrison, 1964).

With the the spacing dependence of CPD ($\pm 0.010\ eV$) and PEYS spectral resolution obtained by measuring half band width of the Hg spectral line at 3131 Å ($\pm 0.046\ eV$), the $q\Phi_R$ value for graphite is estimated to an accuracy within ± 0.05 eV is:

$$q\Phi_R = 4.7 \pm 0.05\ eV$$

There is a good agreement between the present experimental value and the one reported in the literature for colloidal graphite.

Properties such as low intrinsic yield, and uniform and stable work function made this material ideal for its use as a conducting coating in a variety of electron physics apparatus (Taft and Apker, 1955; Feuerbacher and Fitton, 1972; LeClair *et al.*, 1996; Schedin *et al.*, 1998). The value of work function reported in the literature for various forms of graphite such as colloidal graphite coating, polycrystalline graphite and highly ordered pyrolytic graphite (HOPG) span a wide range from 4.40 eV to 4.85 eV. Some of the published work function data are summarized in Table 4.1; it may be seen that the differences can be attributed to the individual nature of the samples, surface conditions, and measurement technique (reference work function used therein). However, except for the two extremes (4.4 and 4.85 V), the work function values in the range 4.6 to 4.7 eV looks reliable for graphite in various forms.

Table 4.1. Work function values published in the literature for various forms of graphite measured using different experimental techniques.

Reference	Form of graphite	Measurement technique	Work function (eV)
Apker *et al.*, 1948	Aquadag	Retarding potential	4.74
Taft and Apker 1955	Polycrystalline	Retarding potential	4.65 to 4.85
Willis *et al.*, 1971	pyrolytic graphite	PEYS	4.7
Burns and Yelke 1969	Aquadag	CPD[*]	4.85
Feuerbacher and Fitton 1972	Polycrystalline Aquadag	PEYS	4.74
Reihl *et al.*, 1986	HOPG	inverse-photoemission	4.7
Ago *et al.*, 1999	HOPG	UPS	4.4
Suzuki *et al.*, 2000	-	UPS	4.6 - 4.7
Hansen and Hansen 2001	HOPG	CPD[**]	4.465
Subrahmanyam and Suresh, 2008 (unpublished)	Aquadag coating	CPD,[*] PEYS	4.71

[*] Gold reference, 4.83 eV.
[**] Saturated calomel electrode reference.

The experiments conducted by Subrahmanyam and Suresh (unpublished) indicate that the work function of colloidal graphite coating $(q\Phi_R)$ after sufficient aging is stable in air ambient and high vacuum. A better estimate of $q\Phi_R$ can be obtained from the combined plot of all the individual qV_{CPD} and $q\Phi_P$ measurements carried out on different metals as shown in Fig. 4.21. The mean value of $q\Phi_R$ estimated from the linear fit with a fixed slope of −1 is,

$$q\Phi_R \text{ for Colloidal graphite} = 4.710 \pm 0.011 \text{ ev}$$

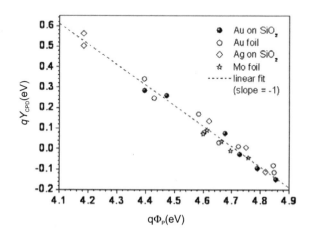

Fig. 4.21. *Comparison of $q\Phi_P$ determined from the PEYS measurements on gold thin film, gold foil, silver thin film and molybdenum foil is plotted against qV_{CPD} measured using Kelvin probe.*

4.6.6 Studies on GaAs (100) Surfaces

As-etched n-GaAs sample is loaded on to the sample holder in the measurement chamber. During the initial stages CPD drift is observed as in the case of metallic samples. It may be noted that the changes in CPD also are contributed by the ambient light induced surface photovoltage decay. PEYS measurements are made after attaining stability in the CPD values under dark conditions. A prolonged evacuation caused appreciable change in CPD and photoelectron emission characteristics. Both CPD and PEYS measurements are repeated after heating the sample to 250°C. Pressure in the chamber during the measurement is $\sim 4 \times 10^{-7}$ mbar. Fig. 4.22 shows the yield versus photon energy plot obtained for various conditions.

Photoemission from semiconductor surfaces prepared by chemical etching is believed to have contribution from filled surface states and valence band emission. Since the density of valence band state is much higher than the surface state density, emission from the valence band dominates the yield curve well above the threshold (Gobeli and Allen, 1965). The onset of yield curve near 4.8 eV (Fig. 4.22) indicates the contribution from filled states in the forbidden energy gap. A quantification of this threshold is rather difficult with the present experimental setup (sensitivity limited).

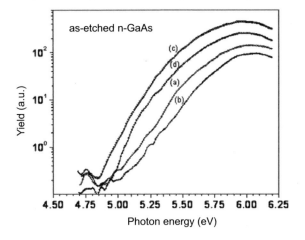

Fig. 4.22. Photoelectric yield in arbitrary units as a function of photon energy. Curves (a) to (d) show data obtained from the same surface of As-etched n-GaAs after, (a) ~ 2 hours of evacuation, (b) prolonged evacuation (more than 24 hours), (c) heated to 250°C and spectra recorded immediately after reaching RT, and (d) ~ 24 hours after heating.

The emission from valence band can be described by non-direct optical excitation in the bulk as shown in Fig. 4.1 and energy loss to quasielastic scattering (Ballantyne, 1972). The ionization energy, "I", is obtained using the power law (equation 4.7); Fig. 4.23 shows the power law fit to the experimental yield curves. The values of I, obtained for the curves (a) to

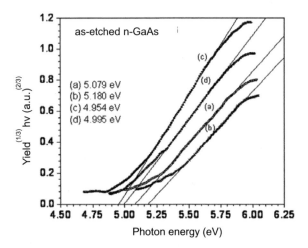

Fig. 4.23. The power law fit to the experimental yield curves of As-etched n-GaAs under various conditions. Curves (a) to (d) show data obtained from the same surface of As-etched n-GaAs after, (a) ~ 2 hours of evacuation, (b) prolonged evacuation (more than 24 hours), (c) heated to 250°C and spectra recorded immediately after reaching RT, and (d) ~ 24 hours after heating.

(e) are given in the Fig. 4.23. The highest value for I (5.18 eV) is observed for as-etched surface, after prolonged evacuation (more than 24 hours).

It is known that the GaAs surface prepared by wet chemical etching in air ambient is covered with several monolayers of oxide (10-50Å) (Shiota et al., 1977). Also, the surface may have deviation from the surface stoichiometry after etching. The ionization energy is very sensitive to the surface dipole layer resulting adsorption and other stoichiometry induced surface structural changes (Hirose et al., 1990). A simple heat treatment (250°C) resulted in a reduction of I as much as ~ 0.185 eV. At these temperatures, desorption of water vapour from the surface and slight changes in arsenic (As) concentration at the oxide interface may occur. Desorption of arsenic requires still higher temperature, above ~ 500°C (Vasquez et al., 1983). For clean GaAs surfaces prepared in UHV, the value of ionization energy reported in the literature lies very close to 5.4 eV (Mönch, 1993; Szuber, 2000).

Using the measured values of qV_{CPD} and I, the position of Fermi level at the surface, E_{FS} and hence the surface band bending, qV_S can be estimated for depletion type surface of n-type semiconductor using the relation,

$$qV_S = q\Phi_S - (I - E_g) - q\phi_b \qquad ...4.9$$

where, the semiconductor surface work function, $q\Phi_S = q\Phi_R - qV_{CPD}$ and the doping dependent bulk potential

$$q\phi_b = 0.026 \ eV \text{ and, } E_g = 1.42 \ eV.$$

Table 4.2 summarizes the values of the ionization energy, CPD, work function, and surface band bending of the as etched GaAs (100) surface observed at various conditions.

Table 4.2. Measured and derived quantities pertinent to the as etched n-GaAs surfaces at various conditions: after, (a) ~ 2 hours of evacuation, (b) prolonged evacuation (more than 24 hours), (c) heated to 250°C and spectra recorded immediately after reaching RT, and (d) ~ 24 hours after heating.

	$I(eV)$ ± 0.05	$qV_{CPD}(eV)$ ± 0.01	$q\Phi_S(eV)$ ± 0.05	$qV_S(eV)$ ± 0.05
(a)	5.079	0.310	4.390	0.705
(b)	5.180	0.228	4.472	0.686
(c)	4.954	0.485	4.215	0.655
(d)	4.995	0.439	4.261	0.66

$$q\Phi_R = 4.7 \pm 0.05 \ eV$$

The surface Fermi level is found to lie between 0.65 to 0.70 eV below the conduction band edge at the surface (E_{CS}) for different conditions. When compared to the changes in the ionization energy (~ 0.2 eV), the surface Fermi-level position changed very little (~ 0.05 eV). The sub-band gap surface photovoltage measurements carried out on these samples revealed the presence of acceptor type filled surface states below the E_{FS}. These results are consistent with the Fermi level pinning close to the middle of the forbidden energy gap reported in the literature (Grant et al., 1981; Yablonovitch et al., 1989; Miller and Stillman, 1990). The minimum surface-state density (Q_{ss}) (acceptor states on n-type) that is required to bring the surface Fermi level, E_{FS} to mid gap can be determined by assuming, $Q_{SS} = -Q_{SC}$. For the energy location of these states close to mid gap, density of surface state required for pinning E_{FS} close to mid gap is found to be slightly larger than intrinsic surface state density (Q_{sc}) of GaAs (100) surface, $3 \times 10^{12} \ cm^{-2}$ which is assumed to be constant at all doping levels (Pashley et al., 1993).

For the observed band bending of ~ 0.70 eV, the space charge layer width may extend up to ~1000 Å into the bulk (for a bulk doping of ~10^{17}/cm³). It should be pointed out that the absorption depth of photons employed for PEYS is only ~100Å ($\alpha > 10^6 \ cm^{-1}$). The expected band bending of 0.1 to 0.2 eV in the region of absorption depth and the changes induced (surface photovoltage) by the exciting photons may influence the emission threshold. The surface photovoltage (SPV) generated above the emission threshold is shown (as hatched region) in Fig. 4.24. It is well-known that

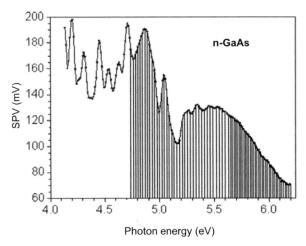

Fig. 4.24. Typical surface photovoltage spectrum of the as-etched n-GaAs sample in the photon energy range of PEYS measurements.

SPV generally leads to reduction in the surface band bending (Kronik and Shapira, 1999). Considering the band flattening and the band bending over the absorption depth, one can conclude that these effects may not cause significant changes in the photothreshold.

The feasibility of measuring the band bending directly using SPV technique will be discussed in Chapter V.

Chapter 5
Surface Photovoltage (SPV) Spectroscopy of Semiconductor Surfaces

5.1 INTRODUCTION

The Surface Photovoltage (SPV) is a powerful technique in evaluating the properties of semiconductor surfaces. This chapter explains the basics and the experimental procedures of Surface Photovoltage technique.

When photons of suitable wavelength interact with the surfaces, electron-hole (e-h) pairs are produced; when these photogenerated electrons and holes are spatially separated, they bring out a change in the surface potential; this change in the surface potential is called the surface photovoltage (SPV). The surface, interface and bulk properties of semiconductors can be evaluated by studying the SPV as a function of (*i*) the incident photon energy, (*ii*) the intensity of illumination, (*iii*) time, (*iv*) temperature and (*v*) strain at the surface. SPV can be measured with a steady state or constant illumination (known as the Direct Current SPV) or a periodic or chopped illumination (Alternating Current SPV).

The SPV signals are generated in several experiments / techniques where the photons interact with the surface, such as photoemission, Raman scattering, photoluminescence, photoreflectance, electron beam induced current (EBIC), and Scanning Tunneling Microscopy (STM). Some of these techniques along with SPV give immense information of the surface of the materials. For example, along with photoemission techniques, surface photovoltage spectroscopy (SPS) has been employed to study the initial stages of Schottky barrier formation on a variety of compound semiconductor surfaces (Brillson, 1975; Brillson, 1982). By combining small signal SPV technique with Field effect and photoconductivity, Brattain and

Garret (Brattain and Garrett, 1956) evaluated the surface band bending, distribution of fast surface states in the energy gap, and their carrier trapping characteristics.

STM and Atomic Force Microscopy (AFM) based approaches have been used to study SPV with high degree of spatial resolution (Hamers *et al.*, 1986; Weaver and Wickramasinghe, 1991). After significant refinements over the years, the SPV method fulfills most of the wafer inspection and basic device characterization demands of the semiconductor industry (Schroder, 2001). SPV is a powerful technique to characterize the 1-D and 2-D quantum structures and grain boundaries of poly crystalline materials (Ashkenasy *et al.*, 1999; Touskova *et al.*, 2002; Dumitras and Riechert, 2003; Shalish, *et al.*, 2000).

5.2 A BRIEF LITERATURE ON SPV TECHNIQUE

SPV technique is identified as a powerful tool right from the early days of investigation of surface properties. Brattain has studied the role of surface states in germanium and silicon through the trapping and recombination of excess carriers generated by band-to-band transition using SPV technique (Brattain, 1947). The surface band bending and the nature of the surface space charge layer have been determined by the saturation behaviour of large signal SPV by Johnson (Johnson, 1958). Most of the investigations carried out on clean and cleaved surfaces of semiconductors employed Kelvin probe method to monitor the CPD and SPV (Brillson, 1977; Liehr and Lüth, 1979; Mönch, 1993). Using sub-band gap illumination, Gatos has studied the energy distribution of surface states using SPV technique (Gatos and Lagowski, 1973; Balestra *et al.*, 1977). Along with STM, the surface photovoltage measurements were carried out to understand the nature and atomic level spatial distribution of defect levels by Aloni *et al.*, (Aloni *et al.*, 2001). Hamers and Cahill demonstrated the capability of SPV effect in STM tunnel junctions for detecting ultrafast light pulse using correlation methods (Hamers and Cahill, 1991). High spatial and depth resolution (ultra-surface-sensitive) SPV measurements have been obtained using near-field illumination techniques (Shikler and Rosenwaks, 2000). The surface photovoltage phenomena has also been used for the detection of infrared radiation (Logothetis *et al.*,1971; Sher *et al.*,1980). Under favourable conditions, the minority carrier life time and diffusion length can be measured from the SPV transient and spectral response measurements (Johnson, 1957; Goodman, 1961). The SPV phenomena and their various applications have been reviewed by Kronik and Shapira (Kronik and Shapira, 1999, 2001). The literature cited here is only indicative.

The SPV technique is to be used with caution; for example, the SPV generated by high energy photons may lead to significant errors in the estimation of Fermi level (Renyu *et al.*, 1987; Baur *et al.*, 1991; Hecht, 1991; Kamada *et al.*, 2000).

5.3 BASICS OF SURFACE PHOTOVOLTAGE SPECTROSCOPY

The absorption of photons of energy equal to or greater than the band gap generates electron-hole pairs by interband transitions (direct or phonon assisted). For a given intensity of illumination, I, the rate of generation of electron-hole pairs, $G(z)$ at a distance z from the illuminated surface can be written as,

$$G(z) = \alpha I (1-R)\eta \exp(-\alpha z) \qquad ...5.1$$

where α, the absorption coefficient, R, the reflection coefficient and η, the quantum efficiency of the semiconductor.

The SPV signal is generated by the spatial separation of the photo generated carriers; the photo generated carriers contribute to the electric field at the surface space charge (Q_{SC}) region or near the surface defect states (Q_{SS}). An externally applied voltage also can contribute to the SPV signal (Many et al., 1965; Kronik and Shapira, 1999, 2001). The electric field induced by the carrier separation, in general, opposes the built-in field; thus there is a reduction in the band bending originally present (prior to illumination), qV_{S0} to a new value qV_S. The surface photovoltage, δV_s is defined as that change in the surface potential,

$$\delta V_S = |V_S - V_{S0}| \qquad ...5.2$$

The equilibrium surface band bending or the surface potential, V_{S0} and its non-equilibrium value, V_S under illumination can be deduced in terms of the total space charge and excess carrier density at the surface ($\triangle p$) using Poisson's equation (Brattain and Bardeen, 1953; Johnson, 1958; Aphek et al., 1998). The essential features of SPV theory proposed by Bardeen (for small signal SPV), Johnson (for large signal SPV) and a recent numerical and analytical model presented by Aphek and coworkers, for wide band gap semiconductors, are outlined in Appendix IV.

The SPV depends mainly on (*i*) the surface potential barrier associated with the surface space charge layer, (*ii*) generation and recombination mechanisms of electrons and holes, (*iii*) density and energy distribution of surface and bulk defect states, and (*iv*) the mobility of electrons and holes. Based on the energy of the exciting photon with reference to the band gap (E_g) of the semiconductor, SPV can be classified as super-band gap ($h\nu > E_g$) and sub-band gap ($h\nu < E_g$). Further, based on the signal strength, it can be categorized into small signal ($SPV \ll kT/q$) and large signal ($SPV \gg kT/q$) methods.

5.4 AN OUTLINE OF THE RELATIONS BETWEEN THE EXCESS CHARGE (Δp) AND SPV

(This section outlines the summary of the Surface Photovoltage and excess carriers. A prior knowledge of Semiconductor Physics is essential to follow the details of the theory. For more details, the reader is requested to consult the references cited).

When a semiconductor (*n* or *p* type) is illuminated with a light of suitable wavelength and known intensity, excess carriers (Δp) are generated; the semiconductor can be in accumulation, depletion and inversion conditions.

The illumination induced excess minority carrier density resulting from the dynamic equilibrium between the rate of generation and of recombination can be obtained from the time independent one dimensional diffusion equation (for *n*-type material),

$$D_p \frac{\partial^2 \Delta p}{\partial z^2} - \frac{\Delta p}{\tau_p} + G(z) = 0 \qquad ...5.3$$

τ_p is the minority carrier life time, D_p is the diffusion coefficient, and $G(z)$ is the generation rate given by equation (5.1). The solution of equation (5.3) can be written (Moss, 1955; Schroder, 2001) with the assumptions: $w \ll L_p \ll d$, $w \ll (1/\alpha) \ll d$, $\Delta p \ll n_0$ where d is the sample thickness, w is space charge layer thickness, as

$$\Delta p = \frac{\eta \varphi (1-R)}{(s^* + D_p / L_p)} \left(\frac{\alpha L_p}{1 + \alpha L_p} \right) \qquad ...5.4$$

Where φ is the photon flux, $L_p = \sqrt{D_p \tau_p}$ is the minority carrier diffusion length and s^* is the effective surface recombination velocity at the illuminated surface (true surface recombination and recombination in the surface space charge region are combined in s^*). It is well known that the recombination rates at the surface are controlled by the surface states. It has been shown (by numerical calculation) that equation 5.4 is applicable for very low values of illumination intensity where the concept of flat quasi-Fermi level is valid (Aphek et al., 1998). In the linear regime of SPV with Δp in the small signal SPV limit, diffusion length of the minority carriers can be determined by measuring the variation of surface photovoltage as a function of optical absorption coefficient (Goodman, 1961; Choo et al., 1993; Schroder, 2001).

The intensity dependant excess charge carrier concentration can also be derived as (numerical solution of the equation given by Aphek et al., 1998).

$$\Delta p = \frac{n_i(F_0^2 - F^2)}{\exp\left(\frac{qV_S}{k_B T}\right) + \exp\left(\frac{-qV_S}{k_B T}\right) - 2} \quad ...5.5$$

F_0 and F denote equilibrium and non-equilibrium space charge functions, respectively. The contribution of surface state charge, Q_{ss} and its change, during illumination is generally treated by Shockley-Read carrier trapping statistics using various models for surface states (Johnson, 1958; Darling, 1991; Aphek et al., 1998). The relevant surface state parameters are described in Appendix (IV). The contribution of Q_{ss} to SPV may be neglected only under special cases (Lile, 1973).

The relationship between SPV and Δp is linear for small signal operation and is very useful for evaluating the surface state parameters (Garrett and Brattain, 1956; Frankl and Ulmer, 1966). For negligibly small surface state density near the Fermi level, the depletion regime small signal SPV for n-type semiconductor can be approximated as (Lile, 1973),

$$\delta V_S = -\frac{kT}{q}\left(\frac{\Delta p}{n_0 n_i^2}\right)\exp\left(\frac{q|V_S|}{k_B T}\right) \quad ...5.6$$

n_0 is the equilibrium majority carrier concentration and n_i is the intrinsic carrier concentration. The equation 5.6 provides an explicit relationship between SPV and surface band bending. Also, it can be seen that SPV increases inversely as n_i^2, implying that the wide band gap semiconductors are very efficient in SPV generation.

The difference in electron and hole capture probability of surface states, r_C and r_V, respectively can give rise to the case of flat band SPV ($V_{S0} = 0$) expressed as (Lile, 1973),

$$\delta V_S = \frac{k_B T}{q}\left(\frac{\Delta p}{n_i}\right)\frac{r_V - r_C}{r_C \exp\left(\frac{(E_F - E_i)_{bulk}}{k_B T}\right) - r_V \exp\left(\frac{-(E_F - E_i)_{bulk}}{k_B T}\right)} \quad ...5.7$$

The flat band signal can either be positive or negative, depending on the ratio r_V/r_C. In the intermediate range of Δp, SPV is a logarithmic function of Δp (Johnson, 1958),

$$\delta V_S = \frac{k_B T}{q}\ln\left(1 + \frac{\Delta p}{p_0}\right) \quad ...5.8$$

This expression can also be derived from the basic transport mechanisms in an illuminated surface barrier (Chiang and Wagner, 1985). In the equation

5.8, the contribution of surface states appear in Δp via the effective surface recombination velocity, s^* (equation 5.4). For very high values of Δp, SPV is expected to saturate to a value of V_{SO}. Under favourable conditions, this typical behaviour of SPV is useful for the determination of V_{SO} (Brillson and Kruger, 1981; Aphek, et al., 1998)

In addition to the surface potential barrier and surface state trapping related SPV, Dember voltage arising from different mobilities of the photogenerated electrons (μ_e) and holes (μ_h) make significant contribution to the total SPV. The Dember voltage, V_D is given by (Goodman, 1961),

$$V_D = \frac{k_B T}{q}\left(\frac{b-1}{b+1}\right)\ln\left(1 + \frac{(b+1)\Delta p}{bn_0 + p_0}\right) \quad ...5.9$$

where b is the ratio of the electron mobility to the hole mobility. Interference of Dember voltage with SPV signal is appreciable for high resistivity samples with larger difference between mobility of electrons and holes.

The SPV theory is versatile to address specific surface and bulk defect states, specific semiconductor materials and specific low dimensional structures: amorphous, 1-D and 2-D quantum structures (Chiang and Wagner, 1985; Choo et al., 1993; Leibovitch et al., 1996; Kronik and Shapira, 1999; Dumitras and Riechert, 2003).

5.4.1 Super-band Gap SPV

When a photon of energy higher than the band gap is incident on a semiconductor, in addition to the regular electron-hole pairs, majority and minority carriers from impurity levels and free carriers associated with the inter band gap states are generated. These carriers also contribute to the SPV signal. Thus the surface state related SPV signal is quite complicated (Gatos and Lagowski, 1973).

5.4.2 Sub-band Gap SPV

When a sub-band gap photon ($h\nu < E_g$) interacts with the semiconductor surface, it generates either electrons or holes from localized surface and bulk electronic states (Chiarotti et al., 1968; Vanmaekelbergh and Pieterson, 1998). This process brings a change in the localized charge in the surface or bulk electronic states. In order to maintain the charge neutrality, a change in the space charge, Q_{SC} will follow. Accordingly, the surface potential (surface barrier height) may change by an amount, δV_s (Fig.5.1). This change is measured as SPV. The sign of the SPV depends on the type of sub-band gap transition and the type of the generated charge. The separation of the contribution of surface states from the bulk states is often very difficult from the experimental and theoretical point of view. However, it may be noted that the sub-band gap SPV inherently is more sensitive to surface states than to bulk states (Leibovitch et al., 1994).

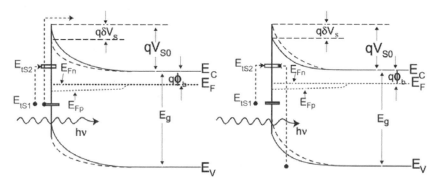

Fig. 5.1. Sub-band gap SPV of n-type semiconductor surface with depletion layer. Empty and filled surface states are shown at E_{tS1} and E_{tS2}, respectively. The quasi Fermi levels are considered flat throughout the depletion layer (a) reduction in barrier height associated with surface state depopulation, (b) increase in barrier height associated with surface state population.

The surface photovoltage associated with electron transition from the filled surface states to the conduction band (depopulation) and from the valence band to the empty surface states (population) provide a unique and convincing method of characterizing both filled and empty surface states (Gatos and Lagowski, 1973).

The spectral distribution of SPV, commonly referred to as surface photovoltage spectra (SPS), gives the energy position of the surface states directly. SPS can be calculated if the parameters of the surface states are known (Szaro et al., 1999). The generation rate (equation (5.1)) of electron hole pairs is important in evaluating the super-band gap SPV. In the sub-band gap SPV, the rates of transition of carriers from surface states into conduction band, G_{tC}^{I} and valence band to surface state, G_{Vt}^{I} are very important.

These rates can be expressed in terms of the photoionization capture cross-section of surface states K_{ph}^{d} and K_{ph}^{p} of depopulation and population transitions, respectively as (Gatos and Lagowski, 1973),

$$G_{tC}^{I} = \phi(hv)K_{ph}^{d}(hv)n_t, \quad G_{Vt}^{I} = \phi(hv)K_{ph}^{p}(hv)p_t \qquad \ldots 5.10$$

$\phi(hv)$ is the photon flux (spatial dependence is neglected owing to the fact that absorption depth of sub-band gap photon is much higher than the depletion layer width and diffusion length) and n_t and p_t are density of electrons and holes in the surface states, respectively. The recombination rate, $R_{Ct,tV}$ and thermal generation rate $G_{tC,Vt}^{T}$ associated with optical generation rate $G_{Ct,Vt}^{I}$ dictate the net change in the surface potential. A quantification of these dynamic parameters associated with the surface states relies on the analysis of SPV transients (Balestra *et al.*, 1977; Kronik and Shapira, 1993).

5.4.3 Principle of SPV Measurement Using Kelvin Probe

The illumination induced change of the band bending, $q\delta V_S$ within the surface space charge region of the semiconductor will appear as the change in the work function of the semiconductor surface. As illustrated in Fig. 5.2, $q\delta V_S$ will appear as the change in CPD, $q\delta V_{CPD}$ between a metallic reference electrode and the semiconductor surface can be measured using a Kelvin probe.

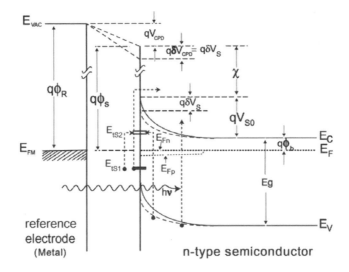

Fig. 5.2. The energy band diagram of metallic reference electrode and an optically illuminated n-type semiconductor with depletion layer at the surface illustrating the SPV measurement principle. Empty and filled surface states are shown at E_{tS1} and E_{tS2}, respectively. It is assumed here that majority carrier quasi Fermi level (E_{Fn}) does not change much with illumination.

Under the non-equilibrium condition, rigorously formulating, one has to consider the quasi-Fermi levels for electrons E_{Fn} and for holes E_{Fp} instead of E_F at the semiconductor surface when illuminated. It is assumed here that majority carrier quasi Fermi level (E_{Fn}) does not change much with illumination (Santos *et al.*, 1999). This phenomenological approximation can be surmounted by considering the fact that δV_S is an *e.m.f.* appearing at the semiconductor surface which can be added or subtracted from the equilibrium CPD value depending on the polarity. This argument also holds good for SPV generated in buried *hetero* or *homo*-junctions and any junction with interfaces within the penetration depth of light and diffusion length of the minority carriers.

5.5 SURFACE PHOTOVOLTAGE EXPERIMENTAL SETUP

A description of the sample holder, high vacuum system, and reference electrode are given in Chapter 3. The optical system for SPS, measurement electronics and data acquisition and controls are illustrated in Fig. 5.3. The photon source for the SPS measurement is a 25 watt tungsten filament lamp. The optics consists of prism monochromator with built-in stray light cutoff filter (Carl Zeiss, M4GII). (Sophisticated monochromators and photon counting systems make the design more elegant and more sensitive). The photon beam is defocused using a front coated concave mirror so as to cover the sample surface slightly in excess of the probe area in a near normal incidence configuration. The sample surface is one electrode of the Kelvin capacitor; the other reference electrode of the Kelvin capacitor is the colloidal graphite coated metal grid; the reference electrode has an optical transparency of about 80% in the working spectrum. For intensity dependant SPV and transient SPV measurements, a variac controlled high power tungsten halogen lamp (150 watt) and band pass filter combination is used in conjunction with the concave mirror. The wave length scan for SPS, voltage control to the tungsten halogen lamp and mechanical shutter are automated using computer controlled stepper motors. The automation is achieved using a Qbasic based computer control program.

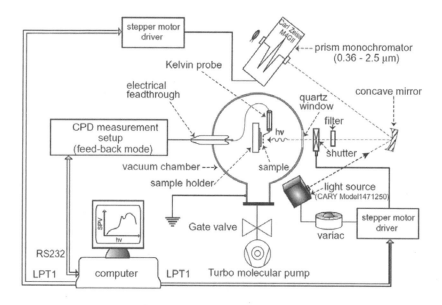

Fig. 5.3. Experiment arrangement of SPV measurement using Kelvin probe.

The photon flux ϕ for the SPS in the energy range 0.49 eV to 2.2 eV is evaluated (Fig. 5.4) by calibrated silicon (PIN 5D, UDT sensors, INC.) and germanium (UDT sensors, INC.) photodiodes operated in the photoconducting configuration with zero bias and a PbS photoconductor operated at 100 volt d.c bias. A cutoff filter with upper energy limit around

2 eV ($\lambda > 620nm$) is introduced in the optical path to minimize the stray light entering into the vacuum chamber. A double side polished GaAs wafer is placed adjacent to the sample to confirm the sub-band gap SPV. The intensity of the high power tungsten halogen lamp with various band pass filters is recorded as a function of control (variac) voltage using the precalibrated detectors. The flux reported here is calibrated and it may vary to within ± 5% due to repositioning of the optical components during sample loading and deloading.

The photon flux, φ, is fairly smooth and has a variation of only an order of magnitude in the sub-band gap regime of GaAs (Fig. 5.4). Because of the logarithmic functional dependence of SPV on φ in the large signal regime (equation 5.6), no attempt has been made to normalize the SPS. The quantitative analysis of measured SPV with theoretical models involving various dynamic parameters of surface and bulk states is beyond the scope of the present book.

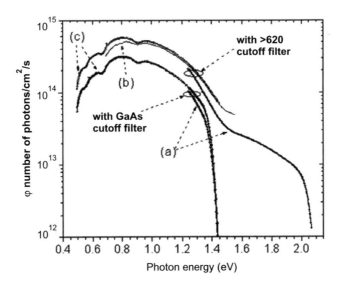

Fig. 5.4. Spectral dependence of photon flux, φ employed for SPS measurements (with GaAs and $\lambda > 620nm$ cutoff filters) measured using, (a) silicon photodiode (PIN 5D. UDT sensors, INC), (b) germanium photodiode (UDT sensors, INC), and (c) PbS photoconductor.

5.5.1 Samples for SPV Measurement

(1) (i) As grown photoconducting thin films of CdTe (111),
   (ii) Boron on vicinal Si(100) (MBE grown, Microphysics Laboratory, University of Illinois at Chicago) and
   (iii) CdS on glass substrate (chemical bath technique) (to demonstrate the band edge SPV effect only).

(2) As etched and aged surfaces of *n*-type Si-doped GaAs (100) wafer with carrier concentration of $\sim 1.7\, 10^{17}\,/cm^3$ (M/s Scientific instruments and materials Inc., USA)

(3) p-type Zn-doped GaAs (100) wafer with carrier concentration of $\sim 8 \times 10^{18}\,/cm^3$ (M/s MTI Corporation, USA)

(4) Semi-insulating (compensated, $\sim 10^{17}/cm^3$-Cr) GaAs(100) (M/s Wacker)

(5) MOCVD grown epitaxial p-GaAs on heavily doped n-GaAs(100) substrate (Material Research Centre, IISc, Bangalore, India).

Samples procured from vendors are subjected to degreasing in trichloroethylene, and acetone in an ultrasonic bath followed by deionized water rinse prior to native oxide etching in dilute HCl. The samples are further subjected to etching in $H_2SO_4: H_2O_2: H_2O$: 7:1:1 for $\sim$ one minute to remove residual contamination and polishing damage. The ohmic contacts for n-type and semi-insulating samples are made by alloying thermally evaporated indium at 350°C for one minute in nitrogen atmosphere. Thermally evaporated gold-germanium alloy is used for p-type samples (alloyed at 450°C for ~1.5 minute in nitrogen ambient).

For the SPV measurements, the Kelvin probe is operated in the feedback loop configuration. The CPD is measured as a function of photon energy (SPS), intensity and time. It may be mentioned that the measurement in the time scale and amplitude are limited by the response time (few seconds) and sensitivity of the Kelvin probe (~ 1mV). The Kelvin method of measuring time dependence of SPV is rather unique among other methods.

5.6 RESULTS

5.6.1 Super-band Gap SPV in GaAs, CdS and CdTe/Silicon Heterojunction

In common semiconductors, it is known that the valence band density of states is much higher than the surface and bulk defect state densities; consequently, very efficient SPV generation mechanism is often associated with absorption of super-band gap photons. Fig. 5.5 illustrates the super-band gap SPV associated with fundamental absorption edge (indicated) of CdS, CdTe, GaAs and Si. Contribution of sub-band gap SPV transitions is also shown in the Fig. 5.5. The sign of SPV in these materials indicates the depletion type of surface space charge layer which is common in compound semiconductor surfaces (Swank, 1967). In the case of CdTe/Si, the buried heterojunction photovoltage is generated by (*i*) band to band transition in silicon and (*ii*) CdTe surface contribution. The dominating SPV signal corresponds to Si and the second threshold is from CdTe surface contribution. The sign of surface and interface SPV signals are found to be the same.

Under favourable conditions (sufficient SPV generation efficiencies), fundamental absorption edge of various forms of semiconductors (powder, thin film, bulk, quantum well, quantum dots, nano) can be identified using the SPV technique (Adamowicz and Szuber, 1991; Kui minski *et al.*, 1991; Gal *et al.*, 1999; Sharma *et al.*, 2002).

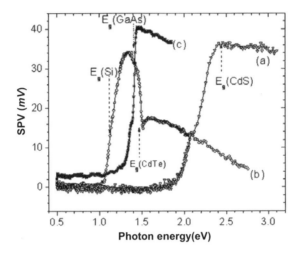

Fig. 5.5. Surface Photovoltage spectra recorded in ambient air using Kelvin probe. (a) CdS, (b) CdTe/Si and (c) well aged n-GaAs (100). Arrows indicate the fundamental absorption edge of these materials.

5.6.2 Moderately Doped n-Type GaAs (100) Surfaces

SPV results presented here for as etched surfaces of *n*-type Si-doped GaAs (100) wafer. The equilibrium value of surface band bending, qV_{so} deduced from CPD and PEYS measurements is found to be close to 0.7 eV for these samples. Accordingly the surface Fermi level position, $E_{FS} = (E_C - E_F)_S$ is also close to 0.7 eV within the experimental error of $\pm 0.05\,eV$. From the knowledge of E_{FS}, one can distinguish the type of surface state SPV transitions (filling or emptying).

SPS Results

Fig. 5.6 shows the SPS plot obtained for various conditions, generated by means of heat treatment in vacuum and prolonged air exposure. The SPS curves are similar in their shapes except for the changes in SPV efficiency. The drop in SPV above *2 eV,* (marked in the Fig. 5.6) is due to the significant reduction in the photon flux (φ with $\lambda > 620\,nm$ cutoff filter shown in Fig. 5.4). The super-band gap SPV associated with fundamental absorption edge is indicated as band gap transition, E_g. The distinct changes in the slope at 1.00 eV, and 1.28 eV, indicates the onset of surface state emptying transitions (depopulation by emission to conduction band) from filled states below E_{FS}. These features can be assigned to the presence of filled states near 1.00 eV and 1.28 eV below the conduction band minimum at the surface, E_{CS}. No prominent surface state filling (population) transition is observed. The decrease of SPV efficiency with aging is evident from curves (a) to (e) in Fig. 5.6.

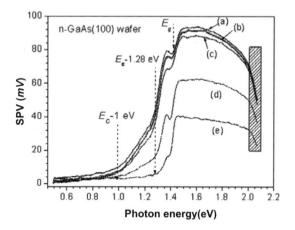

Fig. 5.6 Surface photovoltage spectra of as-etched n-GaAs (100) under various conditions: (a) ~ 1 hour of evacuation, (b) more than 2 hours of evacuation, (c) heated to 250°C and spectra recorded after reaching RT, (d) after a stored in vacuum few day, and (d) well aged after etching spectra recorded in air ambient.

The dip in SPV observed near 1.38 eV peak can be assigned to the surface state filling transition close to E_{CS}. Since this structure is very close to the dominant bulk transition edge where α rises over three orders of magnitude ($10-10^4$ cm^{-1}, Sturge, 1962), it is rather difficult to draw specific conclusion regarding this transition. The value of α shows little variation in the energy range below 1.37 eV (1 to 4 cm^{-1}, Sturge, 1962). No significant difference in the shape has been observed among the spectra recorded in vacuum and air ambient. The SPV increase observed above 1 eV may have possible interference from EL2 defect states which are present in all GaAs bulk crystals (Germanova and Hardalov, 1987; Ikari *et al.*, 1992).

Fig. 5.7 shows the SPS recorded to evaluate the possible influence of stray light from the prism monochromator. The GaAs sample used here is prepared in a different batch. The spectra recorded with silicon (curve a) and GaAs (curve b) cutoff filters shows the surface state transitions near 1 and 1.28 eV.

Curve (d) is added in Fig. 5.7 to show the effect of low temperature ~110 K on SPV. The long SPV settling time (in hours) at these temperatures did not permit reliable measurements.

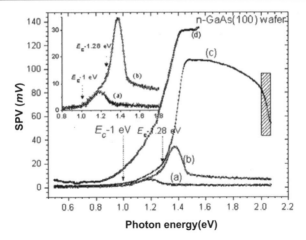

Fig. 5.7. Surface photovoltage spectra of as etched n-GaAs (100) recorded in ambient air with various cutoff filters. (a) double side polished silicon, (b) double side polished GaAs, and (c) $\lambda > 620$ nm built-in cutoff filter of the prism monochromator. (d) spectra recorded at ~ 110 K temperature. Insert in the figure shows the expanded view of the sub-band gap SPV with silicon and GaAs filters.

Intensity Dependence of SPV

The theory of surface photovoltage predicts three regions of dependence on illumination: (*i*) a linear rise in SPV at low level illumination, (*ii*) then a logarithmic dependence for high intensities, and (*iii*) finally, saturation at equilibrium band bending, V_{s0} (Johnson, 1958; Lile, 1973; Aphek *et al.*, 1998; Szaro *et al.*, 1999). Fig. 5.8 shows typical intensity dependence of SPV for various values of super-band gap and sub-band gap photons.

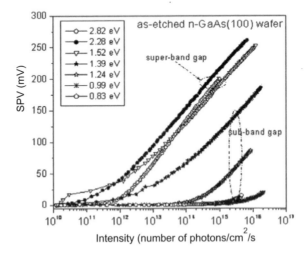

Fig. 5.8. Typical intensity, φ dependence of SPV of as etched n-GaAs (100) measured in vacuum at RT. Labels in the figure correspond to the photon energy

Approximately, three orders of magnitude difference in the relative barrier quantum efficiency of super-band gap and sub-band gap illumination have been observed. The logarithmic dependence of SPV on light intensity described by the equation (5.8) is evident from the super-band gap illumination curves for values of φ above ~ 10^{12} photons/cm^2/s. No sign of SPV saturation is observed over the range of φ used. The sub-band gap SPV also showed the same trend at high photon flux. Complete emptying of the surface states under high illumination is expected to show sub-band gap SPV saturation. The sub-band gap SPV generally requires intensities more than 10^{20} photons/cm^2/s to achieve saturation (Szaro et al., 1999). As shown by Leibovitch and coworkers (Leibovitch et al., 1994), the linear region in the logarithmic scale observed for sub-band gap illumination can be considered as the characteristic of the surface states; where as, the bulk states are expected to show soft saturation at high level illumination.

Fig. 5.9 shows the relation between SPV and excess carrier density, $\triangle p$ calculated using the large signal SPV theory of Johnson (Johnson, 1958; Appendix IV) for the n-GaAs. The relevant bulk parameters used in the calculation are: $n_0 = 1.7 \times 10^{17}\ cm^{-3}$, $n_i = 1.79 \times 10^6\ cm^{-3}$, $\mu_n = 4500\ cm^2/V\text{-}s$, $\mu_p = 150\ cm^2/V\text{-}s$, $L_p = 0.3\ \mu m$, $D_p = 3.9\ cm^2/s$, and $\alpha = 5 \times 10^4\ cm^{-1}$ (at $hv = 2.28\ eV$). The equilibrium surface band bending, $qV_{s0} = 0.7\ eV$ is assumed to result from the discrete surface states of total density, N_{tS} located close to the mid gap at $E_i - E_{tS} = 0.1 eV$. Value of N_{tS} is estimated using the relation, $Q_{SS} = N_{tS} f = Q_{SC}$. The contribution of surface states to SPV is introduced using the surface state parameter χ^2 which is defined as the ratio of hole and electron capture coefficients $\chi^2 = C_p/C_n$.

If χ^2 is large, the surface state captures holes more effectively than the electrons and its charge tends to become more positive upon illumination. Similarly, if χ^2 is small, the surface charge tends to become more negative upon illumination. The "Constant Q_{SC}" curve in Fig. 5.9 is calculated by neglecting the illumination induced change in Q_{SS}. Efficient hole capture corresponds to large χ^2 enhances the SPV efficiency (small values Δp) as shown in Figure 5.9. On the other hand, efficient electron capture corresponds to small χ^2 decreases the SPV efficiency (large values of Δp). The negative surface state charge responsible for the depletion layer at n-type semiconductor surfaces are known to arise from efficient electron capture efficiency of surface states (large χ^2). From the above argument one can infer that the SPV curves for small values of χ^2 describe the real contribution of surface states on n-type semiconductor surfaces.

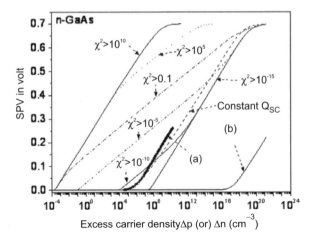

Fig. 5.9. Excess carrier density dependence of SPV calculated using large signal SPV theory of Johnson for n-GaAs sample with $n_0 = 1.7 \times 10^{17}\ cm^{-3}$ and $qV_{S0} = 0.7\ eV$. The value of χ^2 used in the calculation is indicated against each curve using arrows. Experimental SPV curve for $hv = 2.28\ eV$ is shown as curve (a) Curve (b) shows the contribution of Dember potential.

Illumination intensity dependence of Δp or Δn is estimated using the equation (5.4) to compare the experimental curve with the theoretical one. Using the doping dependent effective surface recombination velocity, $s^* = 10^5\ cm^2/s$ for etched surfaces of n-GaAs (Aspnes, 1983) and bulk properties listed above, the conversion factor obtained is $\dfrac{\Delta p}{\phi} = 2.6 \times 10^{-6}\ cm^{-1}s$. This is valid only for very low values of illumination intensity where the assumption of constant s^*, D and L.

The deviation from constant $\Delta p/\varphi$ ratio is shown to be within a factor of two to three over ten orders of change in Δp for moderately doped p-GaAs (Aphek et al., 1998). Using the above mentioned value for, $\Delta p/\varphi$ the intensity dependence of experimental SPV curve is shown in Figure 5.8 for $hv = 2.28\ ev$. It is converted into Δp dependent SPV curve and is shown as curve (a) in Fig. 5.9. A comparison of experimental and theoretical curves indicates the requirement of unrealistically high values of photon flux to achieve SPV saturation for the present case. Curve (b) in Fig. 5.9 shows the contribution of Dember potential calculated using equation (5.9). For n-type samples with depletion layer at the surface, the sign of the Dember potential is same as V_{SO} induced SPV (additive). For the present case, in the intensity range used in the experiment, the Dember potential developed is far below the detection capability of Kelvin probe.

Surface Photovoltage Transient

The time dependence of the SPV signal for various values of super-band gap and sub-band gap photon energies (recorded using the highest values of photon flux shown in the corresponding intensity dependence of SPV curves of Fig. 5.8) is shown in Fig. 5.10. The curves are vertically and horizontally offset for clarity. These SPV transient curves are recorded in vacuum, approximately a day after etching the n-GaAs wafer. A rapid rise is observed irrespective of the energy of the photon used. A rapid initial decay followed by a relatively long tail is observed which is also found to be independent of the photon energy used. It becomes necessary to keep the sample in the dark after each measurement for more than half an hour to reach true dark CPD value. The tails of the decay curve is shown as insert in Fig. 5.10 along with double exponential fit obtained using equation of the form,

$$y = A_1 \exp\left(\frac{-x}{\tau_1}\right) + A_2 \exp\left(\frac{-x}{\tau_2}\right) \qquad ...5.11$$

Typical values obtained are

$$A_1 \sim 25\,mV, \tau_1 \sim 6\,min., A_2 \sim 12\,mV, and\ \tau_1 \sim 0.7\,min.$$

The ratios A_1/A_2 and t_1/t_2 is found to be fairly constant for all the curves.

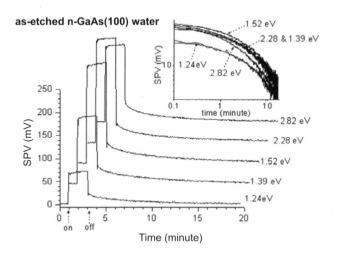

Fig. 5.10. SPV as a function of time for various values of photon energy shown against each curve. On and off time of illumination is indicated with arrows. Tails of the decay curve is shown as insert along with double exponential fit. Sample used is as-etched n-GaAs (100) with $n_0 \sim 1.7 \times 10^{17}\ /cm^3$.

The initial decay is associated with recombination of the excess minority carriers (holes). Long decay tail observed for super-band gap illumination is characteristic of the emission of carriers trapped during illumination in slow surface states (associated with adsorbed over-layer). The emission rates of fast surface and bulk states (in milli seconds or less) are beyond the detection limit of the Kelvin probe. The decay tail observed for sub-band gap illumination can be attributed to the capture rate of slow surface states (Balestra *et al.*, 1977). The surface treatment leads to changes in these time constants. Using the temperature and illumination intensity dependence of decay rate, one can distinguish the various competing mechanisms (Goldstein and Szostak, 1980). Super-band gap SPV transient phenomena in Schottky type surface barrier may also be treated using field emission and thermionic field emission models of current flow (barrier height and width dependent) in open circuit between the surface and bulk (Galbraith and Fischer, 1972; Hecht, 1991).

AC SPV and Diffusion Length of Minority Carriers.

AC SPV generated by chopped illumination is measured using a MIS (metal grid, mica, semiconductor) capacitor geometry (Kronik and Shapira, 1999). A high input impedance FET op-amp (LH0042, National Instruments) in voltage follower configuration and a Lock-in amplifier (EG&G, M5210) together forms the signal detection system. The frequency response of the detection system together with the MIS capacitor is calibrated using square wave of known amplitude (Arbitrary function generator, HP 33120A) (Johnson, 1957). The sensitivity of an AC SPV technique is much higher than that of the DC SPV measurements.

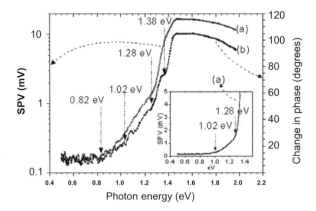

Fig. 5.11. (a) A C SPV (left y-axis in log scale) and (b) change in SPV signal phase (right y-axis) as a function of photon energy for as etched n-GaAs(100) in air ambient. Different vertical scales of the curves are indicated by arrows. Linear scale a.c SPV plot in the sub-band gap photon energy range is shown as insert. Chopping frequency is kept at ~ 11 Hz during the measurement.

Curves (a) and (b) in Fig. 5.11 shows the AC SPV (left y-axis in log scale) and change in the phase of the SPV signal (right y-axis), respectively, measured simultaneously as a function of photon energy for as etched n-GaAs(100) in air ambient. The chopping frequency is kept at ~ 11 Hz during the measurement. The phase of the signal is found to follow the metal-insulator-semiconductor boundaries in SPS. The changes in the slope of sub-band gap SPV observed is found to be similar to that of DC SPS shown in Fig. 5.6, except for the lowest one (0.82 eV). A linear scale SPV plot shown as insert in Fig. 5.11 clearly indicates the dominant changes near 1.02 and 1.28 eV.

At constant frequency, the phase of electrical signal is influenced by the changes in the capacitance, resistance and inductance. Inductance may not be present hence it may not be considered in the present case. During illumination, the resistance of the sample or the capacitance associated with the surface states or depletion region may show appreciable changes. Since the capacitance of the mica spacer (thickness ~10 mm) is much smaller than the depletion layer or the surface state capacitance (thickness < 0.1 mm), the former quantity dominates the total capacitance. With these arguments, one can neglect the contribution of both capacitance and resistance to the observed phase change. Thus, one can infer that the response time of energy states (emission and capture rates for carriers) involved in the SPV generation may be responsible for the observed phase change with photon energy (Liehr and Lüth, 1979). Both the sub-band gap and super-band gap SPV signals are found to be sensitive to chopping frequency.

The Diffusion Length of Minority Carriers

The photon flux, φ required to maintain constant small signal SPV (~ 1 mV and ~ 3 mV) is measured as a function of photon energy (optical absorption coefficient, α (hν)) to determine the minority carrier diffusion length in n-GaAs with $n_0 \sim 1.7 \times 10^{17} \text{cm}^{-3}$ (Goodman's method). Intensity of illumination is adjusted for constant SPV by means of variable transmittance metal grid. Based on the linear relationship between SPV and Δp in the small signal SPV limit (equation 5.4), a plot of φ vs α^{-1}, when extrapolated to $\varphi = 0$, gives a negative intercept along the α^{-1} axis equal to the diffusion length, L_p.

Thickness of the sample (d), depletion layer width (w) and excess minority carrier (Δp) must satisfy the conditions: $w < (1/\alpha) < L_p$, $L_p << d$ and $n_0 >> \Delta p$ for the validity of the diffusion length measurement procedure mentioned above. For the present case, typical values of these parameters are: $d = 350\mu m$, $w = 0.077\,\mu m$ ($for\,qV_{S0} = 0.7\,eV$)

and *range of* $(1/\alpha) = 0.1\,to\,0.3\,\mu m$.

The intensity of illumination is adjusted for small signal SPV to satisfy $\Delta p \ll n_0$.

Figure 5.12 shows the plot of φ vs $(1/\alpha)$ for various values of SPV and chopping frequency. The values of α used here are adapted from Sturge (Sturge, 1962). L_p observed for constant SPV values of ~ 3 mV $L_p = 0.292\,\mu m$, curve (a) and ~ 1 mV $L_p = 0.286\,\mu m$, curve (b) at 225 Hz chopping frequency shows closer agreement.

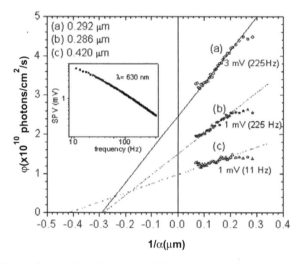

Fig. 5.12. Photon flux, φ plotted against (1/α) for various values of SPV and chopping frequency. Sample : as-etched n-GaAs (100) with $n_0 \sim 1.7 \times 10^{17}\,cm^{-3}$. Values of constant SPV and chopping frequency is shown in curves (a) to (c). Insert in the figure shows the frequency dependence of SPV signal.

The surface effect (contribution of slow states and surface recombination) on the measurement of Lp is evident from the relatively large value obtained at 11 Hz chopping frequency $L_p = 0.420\,\mu m$, curve (c). The observed Lp value is comparable to those reported by other workers for sample doping close to the one used here (Davila and Martinez, 1981).

5.6.3 Heavily Doped p-GaAs (100) Surfaces

In this section, SPV measurements performed on as etched surfaces of heavily doped p-type GaAs (100) wafer (Zn-doped, $p_0 \sim 8 \times 10^{18}\,/cm^3$) are presented (absolute value of the work function is not the main concern here). The heavy doping generally leads to shallow space charge layer with appreciably lower effective barrier height when sufficient amount of surface states are present. As it is shown in equation 5.6, higher doping level generally leads to greater reduction in SPV signal. Thus, applications of SPV method to heavily doped materials are rather limited.

SPS Results

Fig. 5.13 shows the SPS plot obtained for *p*-GaAs under various conditions, generated by means of annealing treatment in vacuum and air exposure. The SPS curves are similar in their shapes except for the changes in the SPV efficiency. The sign of SPV signal indicates the presence of downward band bending at the surface which is typical for *p*-doped semiconductor surfaces. The onset of SPV is observed near the sub-band gap energy 1.32 eV, *i.e.*, about 0.1 eV below the band-gap energy (shown in the inset). In p-type material, the surface photovoltage associated with the sub-band gap illumination induced de-population of surface states is due to emission of holes from the empty surface states to the valence band. Assigning this to the surface states is to be considered tentative because of the heavy doping induced band-gap shrinkage (due to the hole-hole exchange interaction energy) and the corresponding increase of a near 1.32 eV (nearly ten fold increase in α is reported by Casey and coworkers as compared to high pure material (Casey, *et al.*, 1975). A lack of sharp absorption edge observed is consistent with the doping dependent smearing of absorption edge (Casey *et al.*, 1975). The drop in SPV above 2 *eV*, is due to significant reduction in the photon flux (φ with $\lambda > 620\, nm$ cutoff filter, Fig. 5.4).

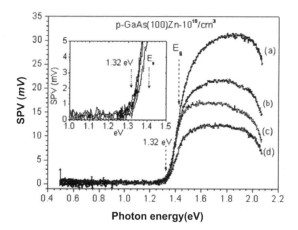

Fig. 5.13. Surface photovoltage spectra of as etched heavily doped ($\sim 8 \times 10^{18} / cm^3$) p-GaAs (100) under various conditions: (a) ~ 2 hours of evacuation, (b) heated to 250°C and spectra recorded after reaching RT, (c) spectra recorded in air ambient after vacuum annealing (d) well aged after etching spectra recorded in air ambient.

Intensity Dependence of SPV

Fig. 5.14 shows the typical intensity dependence of SPV (measured after recording SPS shown as curve (a) in Fig. 5.13) recorded for various values of super-band gap and sub band gap photons. The relatively poor barrier quantum efficiency observed may be attributed to heavy doping, high doping dependent effective surface recombination velocity (Aspnes, 1983) and low

V_{SO}. The shape of 1.52 eV curve shows little saturation behaviour for the highest available photon flux. The comparable efficiencies observed for 1.52 eV, 2.82 eV and 1.39 eV curves show the effect of band gap narrowing (discussed previously for these doping levels).

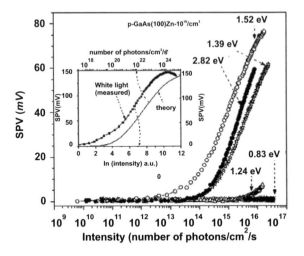

Fig. 5.14. Dependence of the surface photovoltage on the illumination intensity for as etched heavily doped ($\sim 8 \times 10^{18}/cm^3$) p-GaAs (100) in vacuum at RT. Photon energy is indicated against each curve using arrows. Insert shows the saturation behaviour observed using white light (left y-axis, bottom x-axis) and the one calculated using large signal SPV theory of Johnson without considering the effect of surface states, constant Q_{SC} curve (right-y-axis, top x-axis).

The SPV observed for 1.24 eV could be from the contribution of the surface states. The SPV measured with white light is saturated at ~ 150 mV indicating the presence of $qV_{S0} = 0.15\,eV$ (shown as insert of Fig. 5.14). The approximate value of the photon flux required to attain the saturation value of 150 mV calculated using large signal SPV theory of Johnson (constant Q_{SC}, $qV_{S0} = 0.15 eV$) predicts the requirement of very high photon flux (white light flux shown in *a.u.* may be about 10^{19} photons/cm^2/s) as in the case of n-GaAs discussed previously (shown as insert of Fig. 5.14). The value of V_{SO} observed is consistent with the doping dependent movement of Fermi level pinning position reported in the literature for p-GaAs surface with sufficient surface state density (Pashley *et al.*, 1993).

Surface Photovoltage Transient

In Fig. 5.15, the time dependence of the SPV signal is shown for 1.39 eV, 1.52 eV and 2.82 eV photons (recorded using the highest values of photon flux shown in the corresponding intensity dependence of SPV curves of Fig. 5.14). A relatively rapid rise time and long decay time is observed irrespective of the energy of the photon used. The tails of the decay curve is shown as insert in

Fig. 5.15 along with double exponential fit (equation 5.11). The ratios A_1/A_2 and t_1/t_2 is found to be constant for all curves.

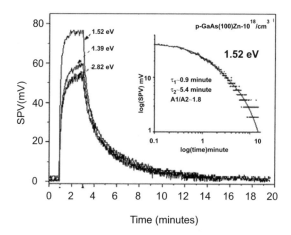

Fig. 5.15. SPV as a function of time for various values of photon energy shown against each curve. Sample used is as-etched heavily doped ($\sim 8 \times 10^{18} / cm^3$) p-GaAs (100). On and off time of illumination is indicated by arrows. Tails of the decay curve is shown as insert along with double exponential fit.

The absence of fast decay component of SPV suggests that the efficient trapping of injected carriers by the slow surface states. The long decay tail observed may be attributed to the subsequent emission of trapped carriers from the slow surface states. An involvement of more than one type of slow states is evident from the distribution of time constants observed.

5.6.4 Semi-insulating GaAs (100) Surfaces

In this section, SPV measurements performed on as etched surfaces of Cr-doped ($\sim 10^{17}$ cm^{-3}) semi-insulating (SI) GaAs (100) sample is presented. The resistivity of the sample is more than $10^7 \, \Omega - cm$. The deep Cr ($Cr^{3+/2+}$) acceptor level close to the middle of the band gap is usually considered to be responsible for the SI characteristics of the sample, also contains EL2 donor states to some extent (Martin, et al., 1980). SI-GaAs is known to exhibit mixed conductivity and comes under the class of relaxation semiconductor (van Roosbroeck and Casey, 1972) and it is widely used as a substrate material for GaAs devices.

SPS Results

Fig. 5.16 shows the surface photovoltage spectra of as etched SI-GaAs(100) under various conditions. A slight decrease in the super-band gap SPV is observed after annealing (curve *b*) and subsequent air exposure (curve *c*). The SPV efficiency is as good as the one observed for moderately doped *n*-GaAs (100).

Four broad SPV regions with distinct changes in slopes close to 1.26 eV, 1.07 eV, 0.75 eV and down to 0.49 eV (indicated in the expanded view) is observed in the sub-band gap region. The photoconductivity of the sample measured with sub-band gap illumination is shown as inset in Fig. 5.16.

The surface Fermi level, E_{FS} is known to be pinned around the mid-gap ($E_{FS} \sim E_F \sim 0.7\,eV$), as a consequence, one can assume flat band at equilibrium even in the presence of surface and bulk defect states (Blakemore, 1982; Liu *et al.*, 1993). The SPV in the super-band gap energy range arise essentially from Dember effect. In the sub-band gap region, various competing carrier generation and trapping mechanisms from both surface and bulk states play an important role (Gatos and Lagowski, 1973; Germanova and Hardalov, 1987).

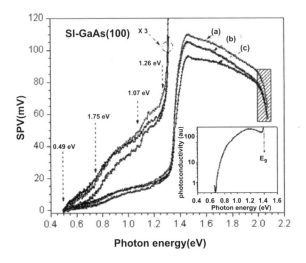

Fig. 5.16. Surface photovoltage spectra of as etched semi-insulating GaAs (100) surface under various conditions: (a) ~ 2 hours of evacuation, (b) heated to 250°C and spectra recorded after reaching RT, (c) spectra recorded in air ambient after vacuum annealing. Expanded view (x 3) of the sub-band gap SPV is shown for clarity. Insert in the figure shows the photoconductivity of the sample in the sub-band gap spectral region.

Generation of carriers of one type (holes or electrons) generally leads to SPV related to the creation of space charge region in the otherwise neutral surface region. The electron-hole pair generation via the intermediary of mid-gap levels (responsible for sub-band gap photoconductivity shown as insert in Fig. 5.16) generates SPV by Dember mechanism as in the case of super-band gap illumination. Since surface and bulk defect states with activation energy ranging from 0.3 eV to 0.8 eV is quite common in Chromium doped SI-GaAs, attributing the SPV features shown in Fig. 5.16 to a specific defect is rather difficult (Kremer *et al.*, 1987; Germanova and Hardalov, 1987; Sharma and Kumar, 2001).

Intensity Dependence of SPV

Fig.5.17 shows typical intensity dependence of SPV measured for various values of super-band gap and sub-band gap photons. A saturation followed by reduction of SPV is observed for near band gap photon energies 1.52 eV and 1.39 eV and a soft saturation for 2.28 eV in the available photon energy range. For photon energy less than 1.39 eV, SPV showed the same trend as that of doped n-GaAs (100) discussed previously. The soft saturation behavior for bulk states at high level illumination suggested by Leibovitch (Leibovitch, et al., 1994) is not observed in the photon flux range used here.

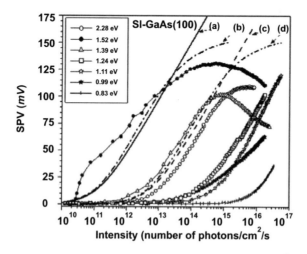

Fig. 5.17. Experimentally observed illumination intensity dependence of SPV (curves with symbols) for the as etched surface of SI-GaAs (Cr-doped) sample. Photon energy correspond to each curve is shown as insert in the figure. Curves (a) and (b) shows the calculated Dember voltage for $s = 3 \times 10^4 / cm^2 / s$ and $s = 2 \times 10^6 / cm^2 / s$, respectively. Curves (c) and (d) are calculated using large signal SPV theory of Johnson (assuming $qV_{S0} = 0.15 eV$, and constant Q_{sc}) for $s = 3 \times 10^4 cm^2 / s$ and $s = 2 \times 10^6 cm^2 / s$, respectively. All the calculated curves assumes 2.28 eV photon.

The Dember voltage calculated using equation (5.9), is shown in Fig. 5.17 as curve (a) and (b) for $s = 3 \times 10^4 / cm^2 / s$ and $s = 2 \times 10^6 / cm^2 / s$ respectively. The SPV calculated using large signal SPV theory of Johnson anticipating weak depletion layer at the surface ($qV_{S0} = 0.15 eV$, and constant Q_{sc}) is shown as Curves (c) and (d) for $s = 3 \times 10^4 cm^2 / s$ and $s = 2 \times 10^6 cm^2 / s$, respectively.

Excess carrier density, Δp is calculated using equation (5.4) by replacing the minority carrier diffusion length and the diffusion coefficient with the corresponding ambipolar quantities relevant to SI-GaAs. The numerical values of bulk parameters used for the estimation of Δp is adapted from the field-assisted photomagnetoelectric experimental results published in the literature for SI-GaAs sample nearly identical to the one used in the present work (Cristoloveanu and Kang, 1984).

Important parameters and their typical values used in the calculation are:

$n_i = 1.79 \times 10^6 \ cm^{-3}$, $\mu_n = 3400 \ cm^2/V\text{-s}$, $\mu_p = 200 \ cm^2/V\text{-s}$,

$\left.\begin{array}{l} \tau_n = 1.1 ns \\ \tau_p = 2.2 ns \end{array}\right\}$ for $s = 3 \times 10^4 / cm^2 / s$, and

$\left.\begin{array}{l} \tau_n = 1.1 ns \\ \tau_p = 1 ns \end{array}\right\}$ for $s = 2 \times 10^6 / cm^2 / s$

For 2.28 eV photon ($\alpha = 5 \times 10^4 \ cm^{-1}$), the above calculation results in the conversion factor, $\frac{\Delta p}{\phi} = 1.22 \times 10^{-5} \ cm^{-1} s$ for $s = 3 \times 10^4 / cm^2 / s$ and

$\frac{\Delta p}{\phi} = 2.9 \times 10^{-7} / cm^{-1} s$ for $s = 2 \times 10^6 / cm^2 / s$.

The calculated curves are found to be very sensitive to the surface recombination velocity assumed (more than two orders of magnitude for the cases considered here). In the case of slight n-type bulk conductivity with upward band bending, the surface space charge field SPV and the Dember voltage will have the same sign and hence the measured SPV is expected to rise after saturation. Opposite is the case with p-type bulk conductivity with downward band bending where one expects reversal in the sign of measured SPV after attaining saturation. From the observed large decrease in SPV for 1.52 eV and 1.39 eV photons, one can safely neglect the contribution of surface space charge field contribution to the measured SPV. Also, based on the above arguments, the possible alternatives of inversion and accumulation may be neglected.

The observed decrease in SPV can be viewed as a result of various other complex competing carrier trapping and emission mechanisms which are known to produce simultaneously SPV's of opposite sign *via* the modification in surface and bulk state charge distribution (Gatos and Lagowski, 1973; Kremer *et al.*, 1987; Liu and Ruda, 1997).

Surface Photovoltage Transient

In Fig. 5.18, the time dependence of the SPV signal for SI GaAs is shown for various values of super-band gap and sub-band gap photon energies (recorded using the highest values of photon flux shown in the corresponding intensity dependence of SPV curves of Fig. 5.17). A rapid rise is observed irrespective of the energy of photon used. The rise and decay of SPV for 1.11 eV and 0.83 eV looks identical in the time scale shown. The negative SPV transient is observed for 1.39 eV, 1.52 eV and 2.28 eV photons followed

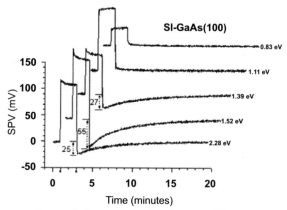

by long recovery time of the order of minutes. The amount of negative swing observed is indicated against each curve. These curves also showed slight decrease in SPV during illumination.

Fig. 5.18. SPV as a function of time for various values of photon energy shown against each curve. Sample used is as-etched Cr-doped SI-GaAs (100). On and off time of illumination is indicated with arrows. The curves are vertically and horizontally offset for clarity.

The amount of negative SPV swing observed for 1.39 eV, 1.52 eV and 2.28 eV transient SPV curves corresponds to the decrease in SPV with illumination intensity shown in Fig. (5.17) for these photon energies. The amplitude of initial negative swing can be considered as the unbalanced trapping related negative space charge SPV superimposed with the genuine **Dember voltage**. The long recovery time observed is found to be similar to persistent photoconductivity and other photoelectric memory effects relevant to the deep traps in SI-GaAs, reported in the literature (Kremer *et al.*, 1987; Jimenez et al., 1987; Vincent *et al.*, 1982; Liu and Ruda, 1997).

5.6.5 GaAs P-N Junction

In this section, SPV measurements performed on as received and etched surfaces of MOCVD grown epitaxial p-GaAs on heavily doped n-GaAs(100) substrate is presented. These measurements are mainly aimed at understanding the effect of buried *p-n* homojunction photovoltage on genuine surface photovoltage signal. The carrier concentration depth profile and the corresponding equilibrium band diagram of the structure are shown in Fig. 5.19. The *p-n* junction is located approximately *1.4 µm* below the surface.

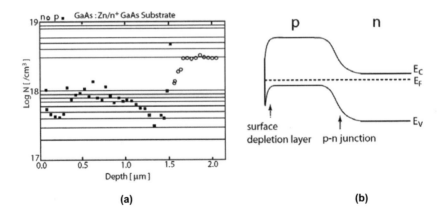

Fig. 5.19. (a) Carrier concentration depth profile obtained from C-V profiling using electrolyte barrier and (b) equilibrium band diagram of the p-n junction (MOCVD grown epitaxial p-GaAs on heavily doped n-GaAs (100) substrate).

SPS Results

Fig. 5.20 shows the SPS plot obtained for as received (curve a) and as etched surfaces (curve b) of MOCVD grown epitaxial p-GaAs on heavily doped n-GaAs (100) substrate measured in air ambient. Two fold increase in SPV signal is observed for super-band gap photon energy after etching. The measured SPV does have contributions from the surface space charge layer (SCL) and buried *p-n* junction SCL. The penetration depth of photons plays an important role in deciding their relative contributions. The sub-band SPV features shown are expected to have their contributions from the surface SCL and *p-n* junction SCL. Since the MOCVD grown homojunctions are known to have far less (negligible) interface state density than the otherwise pinned etched surface of GaAs, the observed sub-band gap feature may be assigned to the surface states in the p-GaAs surface (assuming relatively less SPV efficiency of bulk defects in conducting samples). The sign of SPV signal shows depletion layer at the surface and depopulation of surface states by emission of trapped holes from the empty surface states to valence band with onset near 1.00 eV, 1.23 eV and 1.32 eV. The near band gap SPV will have a dependent contribution from both surface and interface SCL's.

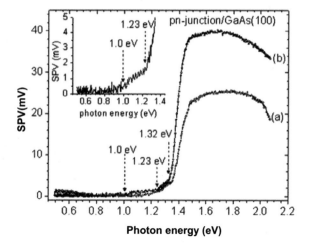

Fig. 5.20. Surface photovoltage spectra of (a) as received and (b) As-etched surfaces of MOCVD grown epitaxial p-GaAs on heavily doped n-GaAs (100) substrate. Expanded view of the sub-band gap SPV is shown as insert for clarity. Spectra recorded in air ambient.

Intensity dependence of SPV

Fig. 5.21 shows the intensity dependence of SPV of as etched surface for various values of super-band gap and sub-band gap photons. A saturation followed by reduction of SPV is observed for near band gap photon energies

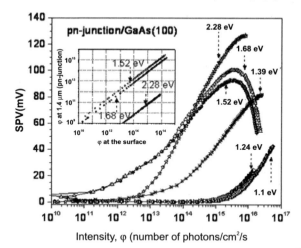

Fig. 5.21. Illumination intensity, φ dependence of the SPV observed for the as etched surface of MOCVD grown epitaxial p-GaAs on heavily doped n-GaAs (100) substrate. Photon energy is indicated against each curve. Insert shows the calculated value of photon flux at the *p-n* junction for the incident photon flux used in the experiment.

1.52 eV and 1.68 eV and a soft saturation for 2.28 eV and 1.39 eV in the available photon energy range. For photon energy less than 1.39 eV, SPV showed the same trend as that of doped *n*-GaAs (100) discussed previously. The observed large decrease for 1.52 eV and 1.68 eV photons indicate the contribution of SPV's of opposite sign from surface SCL and *p-n* junction SCL.

It may be seen (Fig. 5.21) that about 16% of 1.52 eV photons and 6% of 1.68 eV photons incident on the surface are expected to reach the p-n junction. In the case of 2.28 eV photons, only 0.0006% of the incident photons on the surface are expected to reach the *p-n* junction to produce photovoltage of opposite sign. Using accurate numerical values of various quantities in the SPV generation mechanism, one can make a rough estimate of the junction depth and its photovoltage efficiency. Shikler and Rosenwaks demonstrated the effect of buried *p-n* junction in GaP with the near field SPV technique (Shikler and Rosenwaks, 2000).

Surface Photovoltage Transient

In Fig. 5.22, the time dependence of the SPV signal is shown for various values of super-band gap and sub-band gap photon energies for the GaAs p-n junction (recorded using the highest values of photon flux shown in the corresponding intensity dependence of SPV curves of Fig. 5.21.).

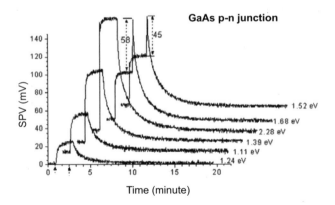

Fig. 5.22. SPV as a function of time for various values of photon energy shown against each curve. On and off time of illumination is indicated with arrows. The curves are vertically and horizontally offset for clarity. Sample used is MOCVD grown epitaxial p-GaAs on heavily doped n-GaAs(100).

Except for the over shoot observed for 1.52 eV and 1.68 eV photons, the raise and decay of SPV is similar to the one shown in Fig. 5.15 for *p*-type surface. The over-shoot observed for 1.52 eV and 1.68 eV photons at the end of illumination period indicates the contribution of relatively fast opposing p-n junction photovoltage. The decay part of the curve shows the surface SCR(Space charge recombination) contribution.

The intensity dependence of SPV transient (Fig. 5.23, curve φ_2) clearly shows the onset of opposing *p-n* junction photovoltage. With increasing intensity, the over shoot component grows at the expense of steady state SPV. Amplitude of overshoot observed within the detection limit of Kelvin probe (response time dependent) agrees fairly well with the corresponding decrease in SPV shown in the intensity dependence curve (shown as insert in Fig. 5.23).

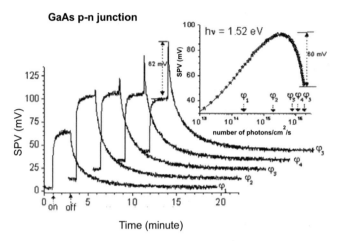

Fig. 5.23. SPV as a function of time for various values of illumination intensity of 1.52 eV photon. Various values of intensity used is indicated in the insert. On and off time of illumination is indicated with arrows. The curves are vertically and horizontally offset for clarity. Sample used is MOCVD grown epitaxial p-GaAs on heavily doped n-GaAs(100).

The observations made here demonstrate the capability of Kelvin probe in resolving photovoltage generated in buried interfaces of semiconductor structures.

■■■

APPENDICES

Appendix I

SURFACE STATES IN SEMICONDUCTORS

A.1 SURFACE STATE MODELS

The Fermi level pinning is observed in most of the semiconductor surfaces during the initial stages of interface formation; this pinning cannot be explained in terms of intrinsic surface states. Several models of surface states have been proposed taking into account, the mechanism of donor or acceptor type defects, their distribution in energy within the forbidden gap and formation of surface or interface dipole layers.

A 1.1 Unified Defect Model

To explain the mechanism of Schottky barrier and metal insulator interface formation, Spicer and co-workers proposed unified defect model (UDM), based on the formation of defect levels with very low coverage of metal or non-metal atoms (one mono layer or less) on vacuum cleaved surfaces of various III-V semiconductors (GaAs, InP, InSb, *etc.*) (Spicer *et al.*, 1980). It is suggested that the surface perturbation results in the production of lattice defects, mainly vacancies of column III or V atoms at or near the interface. The Fermi level pinning position in *n*- and *p*-type III-V semiconductors are attributed to the electron and hole traps associated with columns III and V atom vacancies, respectively. The advanced unified defect model (AUDM) proposed emphasized the role of antisite defect formation during the processing or reactions at interfaces (Spicer *et al.*, 1988). The band energy diagram proposed by Spicer *et al.*, for GaAs is shown in Fig. A1.1. The 0.75 eV and 0.50 eV energy levels predicted by AUDM for As_{Ga} antisite agrees well with photospin resonance results of Weber and Schneider (Weber and Schneider, 1983).

In the case of GaAs, the increase of the ratio of As_{Ga}/Ga_{As} antisites near the surface or interface is expected to move the Fermi level towards the conduction band minimum (CBM); whereas a decrease is expected to move it towards the valence band maximum (VBM). This behaviour accounts well the Fermi level pinning of As rich and Ga rich surfaces obtained during heat treatment and other chemical treatments.

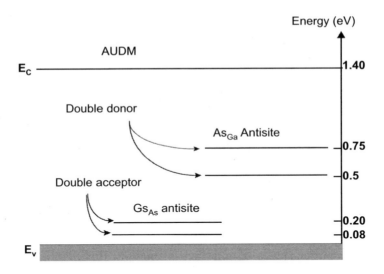

Fig. A.1.1. Energy level diagram for the AUDM of GaAs showing the energy levels of As antisite double donor and compensating Ga antisite double acceptor in the near surface region.

A 1.2 Virtual Gap States of the Complex Band Structure

Virtual gap states (ViGS) of the complex band structure are derived from the bulk band structure of the semiconductor and the electronic states of native defects which are then created at or close to the surface of the semiconductor (Heine, 1965; Chang, 1982). The gap state takes its spectral weight from the local valence and conduction bands, in proportion to its wave function character, and thus exhibit donor and acceptor character respectively. When the Fermi level coincides with the energy at which the virtual gap states crossover from donor to acceptor character, these surface states carry no excess charge. This branch point in the continuous spectrum of the virtual gap states is called the charge neutrality level (CNL) (Mönch, 1993). It lies close to the middle of the average or dielectric gap at the mean-value point as shown in Figure A1.2b. (Mönch,1996). The decay length of the virtual gap states is the shortest at their branch point. There is no discontinuous change in the character of the wave functions at CNL. Rather, states at CNL derive their weight equally from valence and conduction bands. When the effective masses of electrons and holes at the edge of conduction and valence band, respectively, are equal then the branch point of the complex band structure is at the mid gap position. The ViGS around their CNL decreases with increasing width of the band gap. In a one-dimensional model, the density of the ViGS around their CNL was estimated as approximately 4×10^{14} states/cm^2/eV for III-V compound semiconductors.

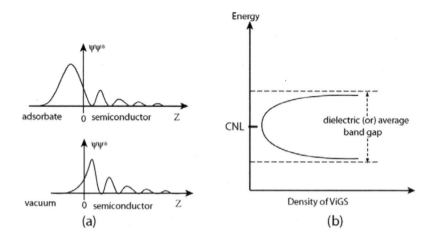

Fig. A.1.2. (a) Wave function of surface states at clean and adsorbate covered surfaces, and (b) Density of ViGS of a linear lattice.

In the band picture, adatom induced surface dipoles are described by the tails of the electron wave functions of the adatoms into the semiconductor as shown in Fig. A1.2a. These tails are derived from the virtual gap states of the complex band structure of the semiconductor. The interplay of the virtual gap states and defect induced electronic states determines the pinning of the Fermi level close to the CNL of the ViGS (Mönch, 1986).

A 1.3 Metal Induced Gap States

The virtual gap states have been explicitly considered to explain the Schottky barrier formation and hetero-junction valence band offsets (Tersoff, 1984a; Tersoff, 1984b; Mönch, 1993). The continuum of states generated within the semiconductor band gap around the Fermi level due to the tailing of wave functions of the metal electrons into the virtual gap states of the complex band structure of the semiconductor is called metal induced gap states (MiGS) (Heine, 1965). The charge transferred between the metal and the semiconductor then pins the Fermi level above, at, or below the CNL of the virtual gap states. The position of Fermi level with respect to CNL of MiGS may be predicted from the knowledge of electronegativity of metal and semiconductor atoms. The general features of MiGS are similar to that of ViGS as discussed in the previous section. The barrier heights which are determined in this way from the bulk band structure of several semiconductors are in agreement with experimental values for interfaces with a variety of metals (Mönch, 1987).

A 1.4 Disorder-induced Gap States

According to the disorder induced gap state (DiGS) model, the position of the surface Fermi level is determined by the charge balance between a U-shaped surface state continuum and the semiconductor bulk (Hasegawa and Ohno, 1986). The DiGS spectrum (Figure A1.3) consists of acceptor-like bonding states and donor like antibonding states separated by a characteristic energy E_{OH} (hybrid orbital energy). The hybrid orbital energy is interpreted as the Fermi energy location of the DiGS spectrum. The position of E_{OH} is independent of the degree of disorder (same even for perfect bonding) as long as the principal bonding scheme of the system remains the same (Hasegawa and Ohno, 1986). The U-shaped curvature and the minimum density at E_{HO} are extremely sensitive to processing. The disordered layer is characterized by the fluctuations of bond lengths and angles due to random stress resulting in bonding mismatch at the surfaces and interfaces. The various surface treatments carried out to modify the disorder, confirms some of the prediction of DiGS model (Hasegawa et al., 1988). The U-shaped distribution with an invariant charge neutrality point (E_{OH}) is a common feature of insulator-semiconductor, metal-semiconductor and hetero-junction interfaces of various compound semiconductors (Hasegawa et al., 1987).

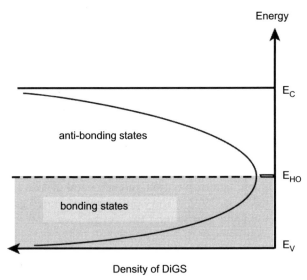

Fig. A.1.3. DiGS model for surface or interface state density distribution.

Several other models based on the effective work function of the reacted interface (Freeouf, and Woodall, 1981) and heat of formation of interfaces (Brillson, 1979) has been proposed to model various types of surface and interface of semiconductors.

Appendix II

The Fourier Coefficients of $C_K(t)$

Fourier series expansion of the periodic function, $C_K(t)$ can be represented by a trigonometric series (Kreyszig, 1999)

$$C_K(t) = a_0 + \sum_{n=1}^{\infty}\left[a_n \cos(n\omega_0 t) + b_n \sin(n\omega_0 t)\right] \qquad ...(A2.1)$$

Here,

$$C_K(t) = \frac{C_0}{1 - m\sin(\omega_0 t)} \quad \text{(equation 3.2)} \qquad ...(A2.2)$$

In terms of simple notations, the above equations can be expressed as,

$$C(\theta) = \frac{1}{1 - m\sin(\theta)} \qquad ...(A2.3)$$

where,

$$\left.\begin{array}{l} \theta = \omega_0 t \\ C(\theta) = \dfrac{C_K(t)}{C_0} \end{array}\right\} \qquad ...(A2.4)$$

Using Euler formula, the coefficients of equation (A1-1) can be written as (Kreyszig, 1999),

$$\left.\begin{array}{l} a_0 = \dfrac{1}{2\pi}\displaystyle\int_0^{2\pi} \dfrac{d\theta}{1 - m\sin\theta} \\[1em] a_n = \dfrac{1}{\pi}\displaystyle\int_0^{2\pi} \dfrac{\cos n\theta}{1 - m\sin\theta}d\theta \\[1em] b_n = \dfrac{1}{\pi}\displaystyle\int_0^{2\pi} \dfrac{\sin n\theta}{1 - m\sin\theta}d\theta \end{array}\right\} \qquad ...(A2.5)$$

Setting $z = e^{i\theta}$, the real integrals in a_0, a_n, and b_n can be evaluated by contour integration.

$$a_0 = \frac{1}{2\pi}\int_0^{2\pi}\frac{d\theta}{1-m\sin\theta} = \frac{1}{2\pi}\frac{-2}{m}\oint\frac{dz}{\left(z^2-\frac{2i}{m}z-1\right)} = \frac{1}{2\pi}\frac{-2}{m}\oint\frac{dz}{(z-\alpha)(z-\beta)}$$

...(A2.6)

where,

$$\left.\begin{array}{l}\alpha = \dfrac{i}{m}\left(1+\sqrt{1+m^2}\right)\\[6pt]\beta = \dfrac{i}{m}\left(1-\sqrt{1+m^2}\right)\end{array}\right\} \text{ for } m<1$$

...(A2.7)

For the simple pole at $z = \beta$, using residue integration method we can write,

$$a_0 = \frac{1}{2\pi}\frac{-2}{m}\oint\frac{dz}{(z-\alpha)(z-\beta)} = \frac{1}{2\pi}\frac{-2}{m}2\pi i\lim_{z\to\beta}\frac{1}{z-\alpha} = \frac{-2i}{m}\frac{1}{\beta-\alpha}$$

...(A2.8)

Using equation (A1.4) and (A1.7), equation (A1.7) can be simplified as,

$$a_0 = \frac{C_0}{\sqrt{1-m^2}}$$ ...(A2.9)

Using the above method, a_n can be written as,

$$a_n = \frac{1}{\pi}\int_0^{2\pi}\frac{\cos n\theta}{1-m\sin\theta}d\theta = \frac{1}{\pi}\frac{1}{m}\oint\frac{z^{2n}+1}{(z-\alpha)(z-\beta)z^n}dz$$ ...(A2.10)

for Ist order pole at $z = \beta$, using residue integration method we can write equation (A2.10) as,

$$\frac{1}{\pi}\frac{1}{m}2\pi i\lim_{z\to\beta}\frac{z^{2n}}{(z-\alpha)z^n} = \frac{2i}{m}\frac{\beta^{2n}+1}{(\beta-\alpha)\beta^n}$$ ...(A2.11)

for n^{th} order pole at $z = o$, using residue integration method we can write equation (A2.10) as,

$$\frac{1}{\pi}\frac{1}{m}2\pi i\frac{1}{(n-1)!}\lim_{z\to 0}\left[\frac{d^{n-1}}{dz^{n-1}}\left(\frac{z^{2n}+1}{(z-\alpha)(z-\beta)}\right)\right] \quad \text{...(A2.12)}$$

Combining the results of equations (A1.11) and (A1.12), one can get,

$$a_n = \frac{2C_0}{m}\sin\left((n-1)\frac{\pi}{2}\right)\frac{m^n}{\left(1+\sqrt{1-m^2}\right)\sqrt{1-m^2}} \quad \text{...(A2.13)}$$

Similarly, b_n coefficient can be obtained by solving,

$$b_n = \frac{1}{\pi}\int_0^{2\pi}\frac{\sin n\theta}{1-m\sin\theta}d\theta = \frac{i}{m}\oint\frac{z^{2n}-1}{(z-\alpha)(z-\beta)z^n}dz \quad \text{...(A2.14)}$$

Extending the residue integration method for I^{st} order pole at $z = \beta$, and n^{th} order pole at $z = 0$, for equation (A1.11) we can get,

$$b_n = \frac{2C_0}{m}\cos\left((n-1)\frac{\pi}{2}\right)\frac{m^n}{\left(1+\sqrt{1-m^2}\right)\sqrt{1-m^2}} \quad \text{...(A2.15)}$$

The Fourier Coefficients of $I(t)$

Current generated by the vibrating capacitor can be written as,

$$I(t) = \frac{d}{dt}\left[\left(-V_{CPD}-V_B\right)C_K(t)\right] \quad \text{...(A2.16)}$$

Using equation (A1.2) for $C_K(t)$ we can write $I(t)$ as,

$$I(t) = \left(-V_{CPD}-V_B\right)m\omega_0 C_0\frac{\cos(\omega_0 t)}{\left(1-m\sin(\omega_0 t)\right)^2} \quad \text{...(A2.17)}$$

In terms of simple notations, the above equations can be expressed as,

$$I(\theta) = \frac{\cos\theta}{\left(1-m\sin\theta\right)^2} \quad \text{...(A2.18)}$$

Where,

$$\left.\begin{array}{l}\theta = \omega_0 t \\ I(\theta) = \dfrac{I(t)}{\left(-V_{CPD}-V_B\right)m\omega_0 C_0}\end{array}\right\} \quad \text{...(A2.19)}$$

Using Euler formula, the Fourier coefficients of equation (A2.18) can be written as,

$$\left. \begin{array}{l} a_0 = \dfrac{1}{2\pi} \displaystyle\int_0^{2\pi} \dfrac{\cos\theta}{(1-m\sin\theta)^2} d\theta = 0 \\[12pt] a_n = \dfrac{1}{\pi} \displaystyle\int_0^{2\pi} \dfrac{\cos\theta}{(1-m\sin\theta)^2} \cos n\theta \, d\theta \\[12pt] b_n = \dfrac{1}{\pi} \displaystyle\int_0^{2\pi} \dfrac{\cos\theta}{(1-m\sin\theta)^2} \sin n\theta \, d\theta \end{array} \right\} \quad \text{...(A2.20)}$$

Setting $z = e$, a_n can be written as,

$$a_n = \frac{1}{\pi} \int_0^{2\pi} \frac{\cos\theta}{(1-m\sin\theta)^2} \cos n\theta \, d\theta = \frac{1}{\pi}\frac{i}{m^2} \oint \frac{(z^2+1)(z^{2n}+1)}{(z-\alpha)^2 (z-\beta)^2 \, z^n} dz$$

...(A2.21)

where,

$$\left. \begin{array}{l} \alpha = \dfrac{i}{m}\left(1+\sqrt{1+m^2}\right) \\[8pt] \beta = \dfrac{i}{m}\left(1-\sqrt{1+m^2}\right) \end{array} \right\} \text{ for } m<1 \quad \text{...(A2.22)}$$

For IInd order pole at $z = \beta$, using residue integration method we can write equation (A2.21) as,

$$\frac{1}{\pi}\frac{i}{m^2} 2\pi i \frac{1}{1!} \lim_{z\to\beta} \frac{d}{dz}\left[\frac{(z^2+1)(z^{2n}+1)}{(z-\alpha)^2 \, z^n}\right] \quad \text{...(A2.23)}$$

for n^{th} order pole at $z = 0$, using residue integration method we can write equation (A2.21) as,

$$\frac{1}{\pi}\frac{i}{m^2} 2\pi i \frac{1}{(n-1)!} \lim_{z\to 0} \frac{d^{n-1}}{dz^{n-1}}\left[\frac{(z^2+1)(z^{2n}+1)}{(z-\alpha)^2 (z-\alpha)^2}\right] \quad \text{...(A2.24)}$$

Combining the results of equations (A2.23) and (A2.24) using simple analytical manipulation, one can get,

Appendix II

$$a_n = \frac{\left(1+(-1)^{n+1}\right)nm^{n-1}}{\left(1+\sqrt{1-m^2}\right)^n \sqrt{1-m^2}} \left[m\omega_0 C_0 \left(V_{CPD} - V_B\right)\right] \quad \ldots(A2.25)$$

Similarly, $b_n = \dfrac{\left(1+(-1)^{n}\right)nm^{n-1}}{\left(1+\sqrt{1-m^2}\right)^n \sqrt{1-m^2}} \left[m\omega_0 C_0 \left(V_{CPD} - V_B\right)\right] \quad \ldots(A2.26)$

■■■

Appendix III

SPECIFICATIONS OF LOW POWER FET-INPUT ELECTROMETER GRADE OP AMP AD515AJ

Suitable for source impedances from $1M\Omega$ to $10^{11}\ \Omega$

Input bias Current: ~ 300 fA

Input noise voltage 0.1 Hz to 10 Hz, $4\mu V\,(p-p)$

  10 Hz $75\text{nV}/\sqrt{\text{Hz}}$

  100 Hz $55\text{nV}/\sqrt{\text{Hz}}$

  1 KHz $50\text{nV}/\sqrt{\text{Hz}}$

Input noise current 0.1 Hz to 10 Hz, $0.007\,pA(p-p)$

  10 Hz $0.01\,pA\ rms$

Input impedance

Differential ($V_{DIFF} = \pm 1V$) $1.6 pF \parallel 10^{13}\Omega$

Common mode $0.8 pF \parallel 10^{15}\Omega$

open-loop gain (d.c), A 40,000 (load resistor greater than 10 $K\Omega$)

open-loop gain (a.c), A @ ~75 Hz 85 dB (load resistor greater than 10 $K\Omega$)

(Note: a.c open loop gain varies linearly from *100 dB* to *60 dB* in the frequency range *10 Hz* to *1KHz*

INPUT IMPEDANCE OF CURRENT PREAMPLIFIER

Input impedance of current preamplifier made using AD515 AJ can be approximated as

$$Z_{in} \approx \frac{R}{1+A}$$

For typical values of $R = 10^9\ \Omega$, and $A = 85\,dB$ @ $75\,Hz$, $Z_{in} \approx 56\ K\Omega$

The impedance of the vibrating capacitor, C_o of few pF or less at w_0 is much higher than Z_{in} of the current preamplifier.

FREQUENCY RESPONSE OF THE CURRENT PREAMPLIFIER

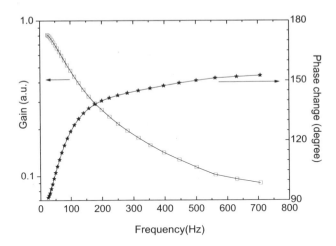

Fig. A3-1. Frequency dependence of gain and phase shift of the current preamplifier.

Appendix IV

THEORY OF LARGE-SIGNAL SURFACE PHOTOVOLTAGE

Johnson's theory of large signal surface photovoltage employed to analyse the intensity dependence of super-band gap SPV results is presented here (Johnson, 1958). The energies and potentials relevant to the theory for a typical n-type semiconductor surface deletion layer are shown in Fig. (A4.1). The surface is represented by the plane at $z = 0$, and the bulk, by positive values of z.

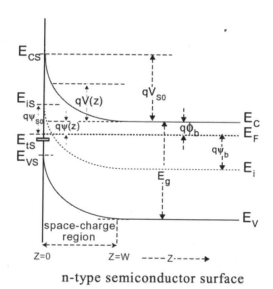

Fig. A4.1. The energy band diagram of n-type semiconductor surface at equilibrium

In the absence of external electric field, the equilibrium surface band-bending or simply, the surface potential, V_{so} is assumed to arise from the charge balance condition of total surface states, Q_{ss} and space-charge, Q_{sc} composed of charge of donors, acceptors, free holes and free electrons.

$$Q_{SS} + Q_{SC} = 0 \qquad \text{...(A4.1)}$$

Q_{SC} can be written as (Johnson, 1957; Many et al. 1965),

$$Q_{sc} = \left(n_i L_D^i\right) F_0 \qquad \text{...(A4.2)}$$

Where $L_D^i = \left(\dfrac{2\varepsilon_0 \varepsilon_S}{n_i k_B T}\right)^{\frac{1}{2}}$ is the intrinsic Debye length ($L_D^i = 2286\ \mu m\ for\ GaAs$) and F_0 is a dimension less quantity called space charge factor. F_0 can be expressed as,

$$F_0 = \pm\left[\Lambda\left(e^{-Y_{S0}} - 1\right) + \Lambda^{-1}\left(e^{Y_{S0}} - 1\right) + \left(\Lambda^{-1} + \Lambda\right)Y_{S0}\right]^{\frac{1}{2}} \qquad \text{...(A4.3)}$$

Here, $\Lambda = \left(\dfrac{p_0}{n_0}\right)^{\frac{1}{2}} = \dfrac{n_i}{n_0} = \dfrac{p_0}{n_i}$, generally referred as doping factor, and Y_{S0} is surface potential at equilibrium in dimensionless form, $Y_{S0} = \dfrac{qV_{S0}}{k_B T}$. Positive and negative signs are used when $Y > 0$ (downward) and $Y < 0$ (upward), respectively.

Under non-equilibrium condition (generated by photons of suitable energy), the space charge factor becomes,

$$F = \pm\left[\Lambda e^P\left(e^{-Y_S} - 1\right) + \Lambda^{-1} e^{-N}\left(e^{Y_S} - 1\right) + \left(\Lambda^{-1} + \Lambda\right)Y_S + \Lambda\left(e^{-Y_S} + e^{Y_S} - 2\right)\Delta_p\right]^{\frac{1}{2}}$$
$$\text{...(A4.4)}$$

where, $P = \beta(E_{FN} - E_F)$, $N = \beta(E_{FP} - E_F)$, $\beta = 1/k_B T$, $\Delta_p = \Delta p/p$ is the fractional change in hole density, $\Delta p = \Delta n$ is the excess carrier density induced by photons. E_{FN} and E_{FP} are quasi Fermi level for electrons and holes, respectively.

In the absence of carrier trapping (constant surface state charge), the introduction of excess carriers can't produce a net charge change in the space charge, instead, will cause charge redistribution. Associated with this is a change in surface potential, ΔY called the surface photovoltage. Based on the assumption that Q_{sc} remain constant during the introduction of excess carriers by photons, we can write,

$$(Q_{SS})_0 = (Q_{SC})_{ill.} = \left(n_i L_D^i\right) F \qquad \text{...(A4.5)}$$

(i.e) $\qquad (Q_{SS})_0 = \left(n_i L_D^i\right) F_0 = \left(n_i L_D^i\right) F \qquad \text{...(A4.6)}$

From this we get,

$$\Delta_p = \frac{F_0^2 - F_Y^2}{\Lambda\left(e^{Y_s} + e^{-Y_s} - 2\right)} \qquad \ldots(A4.7)$$

where $F_Y^2 = \Lambda e^P\left(e^{-Y_s} - 1\right) + \Lambda^{-1}e^{-N}\left(e^{Y_s} - 1\right) + \left(\Lambda^{-1} + \Lambda\right)Y_S$ and $Y_S = Y_0 + \Delta Y$

If the exponential terms dominate in F_0 and F_Y then the surface photovoltage, ΔY will be a logarithmic function of Δp. The relationship between ΔY and Δp will be linear for low level injection.

EFFECT OF SURFACE STATES ON SPV

If the surface state charge, Q_{ss} changes during the introduction of excess carriers by photons then Q_{sc} will also change. For a discrete state E_t having density N_t

$$\left(Q_{SS}\right)_{ill} = N_t\left(f\right)_{ill} = \left(Q_{SC}\right)_{ill} \qquad \ldots(A4.8)$$

The excess carrier dependence of occupancy factor, $(f)_{ill}$ can be written as (Many et al. 1965),

$$(f)_{ill} = \frac{\left(C_n n_S + C_p p_1\right)}{\left[C_n\left(n_S + n_1\right) + C_p\left(p_S + p_1\right)\right]} \qquad \ldots(A4.9)$$

where,

$C_n = N_t v_n \sigma_n$ — capture coefficients of the surface state for electrons
$C_p = N_t v_p \sigma_p$ — capture coefficients of the surface state for holes
$n_1 = n_i e^{-v}$ — equilibrium electron density when $E_t - E_F = 0$
$p_1 = n_i e^{v}$ — equilibrium hole density when $E_t - E_F = 0$
n_S, p_S — densities of electrons and holes at the surface
σ_n, σ_p — capture cross sections for electrons and holes
v_n, v_p — thermal velocities for electrons and holes
$v = (E_t - E_i)/kT$ — Energy level of trap in dimension less form

Using equations (A4.5) and (A4.9), equation (A4.8) can be simplified as (Johnson, 1957),

$$\left[1 + e^{Y - \ln\Lambda + v}\frac{\left(1 + \hat{T}\right)\left(1 + \Lambda^2\Delta_p\right)}{1 + \hat{T}\left(1 + \Lambda^2\Delta_p\right)\left(1 + \Delta_p\right)}\right]^{-1} \quad N_t = \left(n_i L_D^i\right)F \qquad \ldots(A4.10)$$

where, $\hat{T} = \chi^2 e^{2v} e^{-Y+\ln \Lambda - v} / (1 + \Lambda^2 \Delta_p)$ and, $\chi = \left(\dfrac{C_p}{C_n}\right)^{\frac{1}{2}}$

The equation (A4-10) can be solved graphically or numerically to obtain an implicit dependence of surface photovoltage, ΔY on the fractional excess carrier density, Δ_p. The parameter χ^2 determines, in part, the contribution of surface states to the photovoltage.

■■■

References

1. **Adamowicz B and J. Szuber** (1991) Near-band gap transitions in the surface photovoltage spectra for GaAs, GaP, and Si surfaces. *Surface Science*, **247**, 94-99.
2. **Adamowicz, B. and S. Kochowski** (1988) The contribution of surface effects to the surface photovoltage dependence on temperature for the real Si (111) surface. *Surface Science*, **200**, 172-178.
3. **Ago, H., Th. Kugler, F. Cacialli, K. Petritsch, R. H. Friend, W. R. Salaneck, Y. Ono, T. Yamabe, and K. Tanaka** (1999) Work function of Purified and Oxidised Carbon Nanotubes. *Synthetic Metals*, **103**, 2494-2495.
4. **Allen F. G and G. W. Gobeli** (1962) Work function, photoelectric threshold, and surface states of atomically clean silicon. *Physical Review*, **127**, 150-158.
5. **Allen F. G, J. Eisinger, H. D. Hagstrum, and J. T. Law** (1959) Cleaning of silicon surfaces by heating in high vacuum. *Journal of Applied Physics*, **30**, 1563-1571.
6. **Aloni S, Nevo I., and Haase G** (2001). Photovoltage imaging of a single As-vacancy at a GaAs (110) surface: Evidence for electron trapping by a charged defect? *Journal of Chemical Physics*, 115, 1875-1881.
7. **Alves, J. L. A., J. Hebenstreit, and M. Scheffler** (1991) Calculated atomic structures and electronic properties of GaP, InP, GaAs, and InAs(110) surfaces. *Physical Review B*, **44**, 6188-6198.
8. **Amico, A. D, Di Natale, Corrado; Paolesse, Roberto; Mantini, Alessandro; Goletti, Claudio; Davide, Fabrizio; Filosofi, Gabriele** (2000) Chemical sensing materials characterization by Kelvin probe technique. *Sensors and Actuators B: Chemical,* **B70**, 254-262.
9. **Anderson Paul A** (1941) A new technique for preparing monocrystalline metal surfaces for work function study. The work function of Ag (100). *Physical Review*, **59**, 1034-1041.
10. **Anderson, P.A.** (1959) Work function of gold. *Physical Review*, **115**, 553-554.
11. **Apker L, E. Taft, and J. Dickey** (1948) Energy distribution of photoelectrons from polycrystalline tungsten. *Physical Review*, **73**, 46-50.
12. **Apker, L., E. Taft, and J. Dickey** (1948) Photoelectric emission and contact potentials of semiconductors. *Physical Review*, **74**, 1462-1474.

13. **Arakawa, M., S. Kishimoto, and T. Mizutani,** (1997) Kelvin probe force microscopy for potential distribution measurement of cleaved surface of GaAs devices. *Japanese Journal of Applied Physics,* **36(3B)**, 1826-1829.
14. **Arch J. K and S. J. Fonash** (1990) Computer simulation of actual and Kelvin-probe-measured potential profiles: Application to amorphous films. *Journal of Applied Physics,* **68**, *591-600.*
15. **Ashcroft N. W and N. D. Mermin** (1981) Solid State Physics. Holt-Saunders Japan (Ltd), Tokyo.
16. **Ashkenasy N., Leibovitch M., Rosenwaks Y., Shapira Yoram, Barnham K. W. J., Nelson J., and Barnes J.** (1999) GaAs/AlGaAs single quantum well p-i-n structures. A surface photovoltage study. *Journal of Applied Physics,* **86**, 6902-6907.
17. **Aspnes, D. E.** (1983) Recombination at semiconductor surfaces and interfaces. *Surface Science,* **132**, 406-421.
18. **Bacalis, N. C., K. Blathras, P. Thomaides, and D. A. Papaconstantopoulos** (1985) Various approximations made in augmented-plane-wave calculations. *Physical Review B,* **32**, 4849-4856.
19. **Baczyñski J** (1988) Computer-controlled vibrating capacitor technique for determining work function. *Review of Scientific Instruments,* **59**, 2471-2473.
20. **Badwal, S. P. S., T. Bak, S. P. Jiang, J. Love, J. Nowotny, M. R ekas, C. C. Sorrell, and E. E. Vance** (2001) Application of work function measurements for surface monitoring of oxide electrode materials (La, Sr)(Co, Mn, Fe)O_3, *Journal of Physics and Chemistry of Solids,* **62**, 723-729.
21. **Bagus, P. S., Volker Staemmler, and Christof W ll** (2002) Exchange like effects for closed-shell adsorbates: Interface dipole and work function. *Physical Review Letters,* **89**, 096104/1-096104/4.
22. **Baierlein R,** (2001) The elusive chemical potential. American Journal of Physics, 69, 423-434.
23. **Baikie I. D, E. Venderbosch, J. A. Meyer, and P. J. Z. Estrup** (1991) Analysis of stray capacitance in the Kelvin method. *Review of Scientific Instruments*, **62**, 725-735.
24. **Baikie I. D, K. O. van der Werf, and L. J. Hanekamp** (1988) Integrated automatic modular measuring system. *Review of Scientific Instruments,* **59**, 2075-2078.; **Baikie I. D.** (1988) A novel UHV-compatible Kelvin probe and its application in the study of semiconductor surfaces. Ph. D thesis, University of Twente, The Netherlands.
25. **Baikie I. D, K. O. van der Werf, H. Oerbekke, J. Broeze, and A. van Silfhout** (1989) Automatic Kelvin probe compatible with ultrahigh vacuum. *Review of Scientific Instruments,* **60**, 930-934.
26. **Baikie I. D, S. Mackenzie, P. J. Z. Estrup, and J. A. Meyer** (1991) Noise and the Kelvin method. *Review of Scientific Instruments,* **62**, 1326-1332.

27. **Baikie, I. D, Peterman U., Lagel B., and Dirscherl K.** (2001) Study of high- and low-work-function surfaces for hyperthermal surface ionization using an absolute Kelvin probe. *Journal of Vacuum Science & Technology A*, **19**, 1460-1466.
28. **Baikie, I. D.; Petermann, U.; Lagel, B.** (1999) In situ work function study of oxidation and thin film growth on clean surfaces. *Surface Science, 433-435, 770-774.*
29. **Balestra C. L, J. Lagowski, and H. C. Gatos** (1977) Determination of surface state parameters from surface photovoltage transients. *Surface Science*, **64**, 457-464.
30. **Ballantyne, J. M.** (1972) Effect of phonon energy loss on photoemissive yield near threshold. *Physical Review B*, **6**, 1436-1455.
31. **Ban D., E. H. Sargent, St. J. Dixon-Warren, I. Calder, A. J. Spring Thorpe, R. Dworschak, G. Este, and J. K. White** (2002) Direct imaging of the depletion region of an InP p-n junction under bias using scanning voltage microscopy, *Applied Physics Letters*, **81**, 5057-5059.
32. **Bardeen J., and W. H. Brattain** (1948) The transistor, A semiconductor triode. Physical Review, **74**, 230-231.
33. **Bardeen John** (1947) Surface states and rectification at a metal semiconductor contact. *Physical Review*, **71**, 717-727.
34. **Bardeen, J.** (1936), Theory of the work function. II. The surface double layer. *The Physical Review.* **49**, 653-663.
35. **Baumgärtner H and H. D. Liess** (1988) Micro Kelvin probe for local Work function measurements. *Review of Scientific Instruments*, **59**, 802-805.
36. **Baumgartner H.** (1993) A new method for the distance control of a scanning Kelvin microscope. *Measurement Science and Technology*, **3**, 237-238.
37. **Baur, A., M. Prietsch, S. Molodtsov, C. Laubschat, and G. Kaindi** (1991) Surface photovoltage at Cs/GaAs (110): Photoemission experiments and theoretical modeling. *Journal of Vacuum Science & Technology B*, **9**, 2108-2113.
38. **Bellier, J. P.; Lecoeur, J.; Koehler, C.** (1995) Improved Kelvin method for measuring contact potential differences between stepped gold surfaces in ultrahigh vacuum. *Review of Scientific Instruments*, **66**, 5544-5547.
39. **Berglund, C. N. and W. E. Spicer** (1964) Photoemission studies of Copper and Silver: Theory. *Physical Review*, **136**, A1030-A1044.
40. **Bergveld P., J. Hendrikse and W. Olthuis** (1998) Theory and application of the material work function for chemical sensors based on the field effect principle. *Measurement Science and Technology, 9, 1801-1808.*
41. **Besocke K and S. Berger** (1976) Piezoelectric driven Kelvin probe for contact potential difference studies. *Review of Scientific Instruments*, **47**, *840-842.*
42. **Bianconi, A., S. B. M. Hagström, and R. Z. Bachrach** (1977) Photoemission studies of graphite high-energy conduction-band and

valence-band states using soft-x-ray synchrotron radiation excitation. *Physical Review*, **16**, 5543-5548.

43. **Binning G., H. Rohrer, Ch. Gerber, and E. Weibel** (1982) Surface studies by scanning tunneling microscopy. *Physical Review Letters*, **49**, 57-61.
44. **Blakemore, J. S.** (1982) Intrinsic density $n_i(T)$ in GaAs: Deduced from band gap and effective mass parameters and derived independently from Cr acceptor capture and emission coefficients. *Journal of Applied Physics*, **53**, 520-531.
45. **Blott B. H and T. J. Lee** (1969) A two frequency vibrating capacitor method for contact potential difference measurements. *Journal of Physics E: Scientific Instruments*, **2**, 785-788.
46. **Bonnet J., J. M. Palau, L. Soonckindt, and L. Lassabatere** (1977) On the interest of a current amplifier associated with a Kelvin vibrating capacitor. *Journal of Physics E: Scientific Instruments*, **10**, 212-213.
47. **Brattain W. H** (1947) Evidence for surface states on semiconductors from change in contact potential on illumination. *Physical Review*, **72**, 345.
48. **Brattain W. H and C. G. B. Garrett** (1956) Combined measurements of field effect, surface photo-voltage and photoconductivity. *The Bell System Technical Journal*, **35**, 1019-1040.
49. **Brattain W. H., and John Bardeen** (1953) Surface properties of Germanium. *The Bell System Technical Journal*, **32**, 1-41.
50. **Brillson L. J** (1975) Observation of extrinsic surface states on (1120) CdS. *Surface Science*, **51**, 45-60.
51. **Brillson L. J** (1975) Surface photovoltage and Auger spectroscopy studies of (1120) CdS surface. *Journal of Vacuum Science and Technology*, **12**, 249-252.
52. **Brillson L. J** (1977) Surface electronic and chemical structure of (1120) CdSe: comparison with CdS. *Surface Science*, **69**, 62-84.
53. **Brillson L. J** (1982a) Chemical and electronic structure of compound semiconductor-metal interfaces. *Journal of Vacuum Science and Technology*, **20**, 652-658.
54. **Brillson L. J** (1982b) *Surface photovoltage measurements and Fermi level pinning: comments on "Development and confirmation of the unified model for Schottky barrier formation and MOS interface states on III-V compounds. Thin Solid Films*, **89**, L27-L33.
55. **Brillson L. J** (1982c) The structure and properties of metal-semiconductor interfaces. *Surface Science Reports*, **2**, 123-326.
56. **Brillson L. J** (1983) Advances in understanding metal-semiconductor interfaces by surface science techniques. *Journal of Physics and Chemistry of solids*, **44**, 703-733.
57. **Brillson L. J and C. H. Griffiths** (1978) Surface photovoltage spectroscopy of defects and impurities in trigonal selenium. *Journal of Vacuum Science and Technology*, **15**, 529-532.

58. **Brillson L. J and D. W. Kruger** (1981) Photovoltage saturation and recombination at Al-GaAs interfacial layer. *Surface Science*, **102**, 518-526.
59. **Brodie, I.** (1995) Uncertainty, topography, and work function. *Physical Review B*, **51**, 13660-1313668.
60. **Bruening Merlin, Ellen Moons, David Cahen and Abraham Shanzer** (1995) Controlling the work function of CdSe by Chemisorption of Benzoic acid derivatives and chemical etching, *Journal of Physics and Chemistry of solids*, **99**, 8368-8373.
61. **Buczkowski A, G. Rozgonyi, F. Shimura, and K. Mishra** (1993) Photoconductance minority carrier lifetime vs. surface photovoltage Diffusion length in silicon. *Journal of Electrochemical Society*, **140**, 3240-3245.
62. **Burns Jay and Edward Yelke** (1969) Work function of conductive coatings on glass. *Review of Scientific Instruments*, **40**, 1236-1237.
63. **Burstein L, J. Bregman, and Yoram Shapira** (1990) Surface photovoltage spectroscopy of gap states at GaAs and InP metal interfaces. *Applied Physics Letters*, **57**, 2466-2468.
64. **Burstein L, J. Bregman, and Yoram Shapira** (1991) Characterization of interface states at III-V compound semiconductor-metal interfaces. *Journal of Applied Physics*, **69**, 2312-2316.
65. **Butz R and H. Wagner** (1977) A device for measuring contact potential differences with high spatial resolution. *Applied Physics*, **13**, 37-42.
66. **Byun Y and B. W. Wessels** (1988) Surface photovoltage spectroscopy of surface states on indium phosphide. Applied *Physics Letters*, **52**, 1352-1354.
67. **Casey Jr., H. C., D. O. Sell, and K. W. Wecht** (1975) Concentration dependence of the absorption coefficient for and p-type GaAs between 1.3 and 1.6 eV. *Journal of Applied Physics*, **46**, 250-257.
68. **Chan C., W. Gao, and A. Kahn** (2004) Contact potential difference measurements of doped organic molecular thin films. *Journal of Vacuum Science and Technology A*, **22**, 1488-1492.
69. **Chaney J. A., and Pehr E Pehrsson** (2001) Work function changes and surface chemistry of oxygen, hydrogen, and carbon on indium tin oxide. *Applied Surface Science*, **180**, 214-226.
70. **Chang S, I. M. Vitomirov, L. J. Brillson, D. F. Rioux, P. D. Kirchner, G. D. Pettit, J. M. Woodall, and M. H. Hecht (1990) Confirmation of the temperature-dependent photovoltaic effect on Fermi-level measurements by photoemission spectroscopy.** *Physical Review B*, **41**, *12299-12302.*
71. **Chelikowsky, J. R.** and **M. L. Cohen** (1976) Nonlocal pseudopotential calculations for the electronic structure of eleven diamond and zinc-blende semiconductors. *Physical Review B*, **14**, 556-582
72. **Chiang C. L and S. Wagner** (1985) On the theoretical basis of the surface photovoltage. *IEEE Transactions on Electron Devices*, **ED-32**, 1722–726.

73. **Chiarotti, G, G. Del Signore, and S. Nannarone** (1968) Optical Detection of surface states on cleaved (111) surfaces of Ge. *Physical Review Letters*, **21**, 1170-1172.
74. **Chiarotti G, S. Nannarone, R. Pastore, and P. Chiaradia** (1971) Optical absorption of surface states in ultrahigh vacuum cleaved (111) surfaces Ge and Si. *Physical Review* **B**, 4, 3398-3402.
75. **Choo S. C** (1995) Theory of surface photovoltage in a semiconductor with Schottky contact. *Solid-State Electronics*, **38**, 2085-2093.
76. **Choo S. C, L. S. Tan and H. H. See** (1993) Theory of surface photovoltage in a semiconductor with deep impurities. *Solid-State Electronics*, **36**, 989-999.
77. **Choo S. C, L. S. Tan, and K. B. Quek** (1992) Theory of the photovoltage at semiconductor surfaces and its application to diffusion length measurements. *Solid-State Electronics*, **35**, 269-283.
78. **Chung Y. S., Keenan Evans, and William Glaunsinger** (1998) Work function response of thin gold film surfaces to phosphine and arsine. *Applied Surface Science*, **125**, 65-72.
79. **Clabes. J. and Henzler** (1980) Determination of surface states on Si(111) by surface photovoltage spectroscopy. *Physical Review B*, **21**, 625-631.
80. **Clewley, J. D, A. D. Crowell, and D. W. Juenker** (1971) Changes in Photoelectron emission from molybdenum due to exposure to gases. *Journal of Vacuum Science and Technology*, **9**, 877-881.
81. **Cohen M. L, and J. C. Phillips (1965) Spectral analysis of photoemissive yields in Si, Ge, GaAs, GaSb, InAs, and InSb.** Physical *Review,* **139**, A912-A920.
82. **Compton K T, and Irving Langmuir** (1930) Electrical discharges in gases: Part I. Survey of fundamental processes. *Reviews of Modern Physics*, **2**, 123-242.
83. **Comsa G, A. Gelberg, and B. Iosifescu** (1961) Temperature dependence of the work function of Metals (Mo, Ni). *Physical. Review,* **122**, 1091-1100.
84. **Craig P. P, and Veljko Radeka** (1970) Stress dependence of contact potential: The ac Kelvin method. . *Review of Scientific Instruments*, **41**, 258-264.
85. **Cutler P. H and J. C. Davis** (1964) Reflection and transmission of electrons through surface potential barriers. *Surface Science*, **1**, 194-212.
86. **Czanderna Alvin Warren**, Methods of Surface Analysis, Elsevier Science & Technology 1989.
87. **Danyluk S** (1972) A UHV guarded Kelvin probe. *Journal of Physics E: Scientific Instruments*, **5**, 478-480.
88. **Datta S., Gokhale M. R., Shah A. P., Arora, B. M., and Kumar Shailendra** (2000) Temperature dependence of surface photovoltage of bulk semiconductors and the effect of surface passivation. *Applied Physics Letters*, **77**, 4383-4385.

89. **Darling, R. B.** (1991) Defect-state occupation, Fermi-level pinning, and illumination effects on free semiconductor surfaces. *Physical Review B,* **43**, 4071-4083.
90. **Davila, J. D.** and **A. L. Martinez** (1981) Measurements of submicron hole diffusion lengths in GaAs by a photovoltaic Technique. *Applied Physics Letters,* **38**, 442-444.
91. **Demuth J. E, W. J. Thompson, N. J. DiNardo, and R. Imbihl** (1986) Photoemission-based photovoltage probe of semiconductor surface and interface electronic structure. *Physical Review Letters,* **56,** 1408-1411.
92. **Dirscherl k., Iain Baikie, Gregor Forsyth, and Arvid van der Heide** (2003) Utilization of a micro-tip scanning Kelvin probe for non-invasive surface potential mapping of a mc-Si solar cells. *Solar Energy Materials & Solar Cells,* **79**, 485-494.
93. **Dumitras Gh. and Riechert H.** (2003) Determination of band offsets in semiconductor quantum well structures using surface photovoltage. *Journal of Applied Physics,* **94**, 3955-3959.
94. **Eastman, D. E.** (1970) Photoelectric Work Functions of Transition, Rare-Earth, and Noble Metals. *Physical Review B,* **2**, 1-2.
95. **Endriz J. C., W. E. Spicer** (1971) Study of aluminum films. II. Photoemission studies of surface-plasmon oscillations on controlled-roughness films. *Physical Review B,* **12**, 4159-4184.
96. **Ewing Joan R and Lloyd P. Hunter** (1975) A study of the surface *photovoltage of silicon. Solid-State Electronics, 18, 587-591.*
97. **Fain Jr. S.C, and J. M. McDavid** (1974) Work function variation with alloy composition: *Ag-Au. Physical Review B,* **9**, *5099-5107.*
98. **Fain S. C, Jr., L. V. Corbin, II, and J. M. McDavid** (1976) Electrostatically driven apparatus for measuring work function differences. *Review of Scientific Instruments,* **47**, 345-347.
99. **Fall, C. J., N.Binggeli, and A. Baldreschi** (2001) Theoretical maps of work function anisotropies. *Physical. Review,* **B 65**, 045401 (1-4).
100. **Fan H. Y.,** (1942) Theory of electrical contact between solids. *Physical Review,* **61**, 365-371.
101. **Farnsworth H. E and Ralph P. Winch** (1940) Photoelectric work function of (100) and (111) faces of silver single crystals and their contact potential difference. *Physical Review,* **58**, 812-819.
102. **Feibelman, P. J., and D. E. Eastman** (1974) Photoemission spectroscopy-correspondence between quantum theory and experimental phenomenology. *Physical Review B,* **10**, 4932-4947.
103. **Feuerbacher, B. and B. Fitton** (1972) Experimental Investigation of Photoemission from Satellite Surface Materials. *Journal of Applied Physics,* **43**, 1563-1572.
104. **Feuerbacher B, Fitton B, and Willis R. F** (edit.) (1978), Photoemission and the electronic properties of surfaces, John Wiley & Sons, New York.

105. **Fischer, T. E, and P. E. Viljoen** (1971) Electric-field dependence of the surface dipoles of semiconductors. *Physical Review Letters,* **26**, 549-551.
106. **Forget, L.; Wilwers, F.; Delhalle, J.; Mekhalif, Z.** (2003) Surface modification of aluminum by n-pentylphosphonic acid acid: XPS and electrochemical evaluation. *Applied Surface Science,* **205**, 44-55.
107. **Fowler, R. H.** (1931) The analysis of photoelectric sensitivity curves for clean metals at various temperatures. *Physical Review,* **38**, 45-56.
108. **Frankl D. R and E. A. Ulmer** (1966) Theorey of small-signal photovoltage at semiconductor surfaces. *Surface Science,* **6**, 115-123.
109. **Freeouf, J. L., and J. M. Woodall** (1981) Schottky barriers: An effective work function model. *Applied Physics Letters,* 39, 727-729.
110. **Fujihira Masamichi** (1999) Kelvin probe force microscopy of molecular surfaces. *Annual Review of Material Science,* **29**, 353-380.
111. **Gal D., Mastai Y., Hodes G., and Kronik L.** (1999) Band-gap determination of semiconductor powders via surface photovoltage spectroscopy. *Journal of Applied Physics,* **86**, 5573-5577.
112. **Galbraith L. K and T. E. Fischer** (1972) Temperature- and illumination-dependence of the work function of gallium arsenide. *Surface Science,* **30**, 185-206.
113. **Garrett C. G. B and W. H. Brattain** (1956) Distribution and cross-sections of fast states on germanium surfaces. The Bell System Technical Journal, 35, 1041-1058.
114. **Gatos Harry C and Jacek Lagowski** (1973) Surface photovoltage spectroscopy - A new approach to the study of high-gap semiconductor surfaces. *Journal of Vacuum Science and Technology,* **10**, 130-135.
115. **Germanova K, Ch Hardalov, V. Strashilov, and B. Georgiev** (1987) An improved apparatus for surface photovoltage studies with a bimorphous piezoelectric Kelvin probe. *Journal of Physics E,* **20**, 273-276.
116. **Germanova, K.** and **Ch. Hardalov** (1987) EL2 Deep Level in Sub-Bandgap Surface Photovoltage Spectra in GaAs Bulk Crystals. *Applied Physics A,* **43**, 117-121.
117. **Germanova K, L. Nikolov, and Ch. Hardalov** (1989) High sensitive automated setup for measuring surface photovoltage spectra. *Review of Scientific Instruments,* **60**, 746-748.
118. **Glatzel, Th.; Sadewasser, S.; Shikler, R.; Rosenwaks, Y.; Lux-Steiner, M. Ch.** (2003) Kelvin probe force microscopy on III-V semiconductors: the effect of surface defects on the local work function. *Materials Science & Engineering, B: Solid-State Materials for Advanced Technology,* B**102**, 138-142.
119. **Gobeli, G.W. and F.G. Allen** (1965) Photoelectric properties of cleaved GaAs, GaSb, InAs, and InSb surfaces; Comparison with Si and Ge. *Physical Review,* **137**, A245-A254.

120. **Goldstein Bernard and Daniel J. Szostak** (1980) Surface photovoltage, band-bending and surface states on a-Si:H. *Surface Science*, **99**, 235-258.
121. **Goldstein Bernard, David Redfield, Daniel J. Szostak, Laster A. Carr** (1981) Electrical characterization of solar cells by surface photovoltage. *Applied Physics Letters*, **39**, 258-260.
122. **Goodman A. M.** (1961) A method for the measurement of short minority carrier diffusion lengths in semiconductors. *Journal of Applied Physics*, **32**, 2550-2552.
123. **Guichar, G. M.; Balkanski, M.; Sebenne, C. A.**(1979) Semiconductor surface state spectroscopy. *Surface Science*, **86**, 874-87.
124. **Gyftopoulos E. P., and G. N. Hatsopoulos (1968)** Quantum-thermodynamic definition of electronegativity. *Proceedings of the National Academy of Sciences of the United States of America*, **60**, 786-793.
125. **Hadjadj, A.; Roca I Cabarrocas, P.; Equer, B.** (1995) Analytical compensation of stray capacitance effect in Kelvin probe measurements. *Review of Scientific Instruments*, 66, 5272-6.
126. **Halas, S.** and **T. Durakiewicz** (1998) Work functions of elements expressed in terms of the Fermi energy and the density of free electrons. *Journal of Physics: Condensed Matter,* **10**, 10815-10826.
127. **Hansen, W. N,** and **G. J. Hansen** (2001) Standard reference surfaces for work function measurements in air. *Surface Science*, **481**, 172-184.
128. **Hansen, W. N.** and **K. B. Johnson** (1994) Work function measurements in gas ambient. *Surface Science*, **316**, 373-382.
129. **Hamers, R. J.** and **D. G. Cahill** (1991) Ultrafast time resolution in scanned probe microscopies: Surface photovoltage on Si(111)-(7×7). *Journal Vacuum Science and Technology B*, **9**, 514- 518
130. **Hamers R. J., R. M. Tromp, and J. E. Demuth** (1986) Surface electronic structure of Si (111)-(7´7) resolved in real space. *Physical Review Letters*, **56**, 1972-1975.
131. **Han L. T, and F. Mansfeld** (1997) Scanning Kelvin probe analysis of welded stainless steel. *Corrosion Science,* **39**, 199-202).
132. **Hannay N. B., and C. P. Smyth** (1946) The dipole moment of hydrogen fluoride and the ionic character of bonds. *Journal of American Chemical Society*, **68**, 171-173.
133. **Hansen, W. N,** and **G. J. Hansen** (2001) Standard reference surfaces for work function measurements in air. *Surface Science*, **481**, 172-184.
134. **Harris L. B and J. Fiasson** (1984) Vibrating capacitor measurement of surface charge. *Journal of Physics E: Scientific Instruments*, **17**, 788-792.
135. **Hayashi N., H. Ishii, Y. Ouchi and K. Seki** (2002) Examination of band bending at buckministerfullerene (C60)/metal interfaces by the Kelvin probe method. *Journal of Applied Physics*, **92**, 3784-3793.

136. Hecht M. H (1990) Role of photocurrent in low-temperature photoemission studies of Schottky-barrier formation. *Physical Review B*, **41**, 7918-7921.
137. Hecht, M. H. (1991) Time dependence of photovoltaic shifts in photoelectron spectroscopy of semiconductors.. *Physical Review B*, **43**, 12102-12105
138. **Heiland G and W. Mönch (1973) Surface studies by modulation spectroscopy.** *Surface Science,* **37**, 30-47.
139. Henk, J., W. Schattke, H. Carstensen, R. Manzke, and M. Skibowski (1993) Surface-barrier and polarization effects in the photoemission from GaAs (110). Physical Review B, **47**, 2251-2264.
140. Herring C and M. H. Nichols (1949) Thermionic emission. Reviews of *Modern Physics,* **21**, 185-270.
141. Hirose, E., Foxman, T. Noguchi, and M. Uda (1990) Ionization-energy dependence on GaAs (001) surface superstructure measured by photoemission-yield spectroscopy. *Physical Review B*, **41**, 6076-6078.
142. Hölzl J and P. Schrammen (1974) A new pendulum device to measure contact potential differences. *Applied Physics*, **3**, 353-357.
143. Hölzl, J. *Solid Surface Physics.* pp 86-95, Springer, Berlin, 1979.
144. **Holscher AA (1966)** A field emission retarding potential method for measuring work functions, Surf Sci **4**, 89-102.
145. **Hwang I. G and D. K. Schroder** (1993) Effect of wafer stress on surface photovoltage diffusion length measurements. *Solid-State Electronics,* **36**, 1147-1153.
146. **Ikari, T., A. Fukuyama, K. Maeda, and K. Futagami, S. Shigetomi, and Y. Akashi** (1992) Photoacoustic signals of n-type GaAs layers grown by molecular-beam epitaxy on semi-insulating substrates. *Physical Review B*, **46**, 10173-10178.
147. **Ing-Shin Chen, T. N. Jackson, and C. R. Wronski** (1996) Characterization of semiconductor heterojunctions using internal photoemission. *Journal of Applied Physics,* **79**, 8470-8474.
148. **Inoue, N., T. Higashino, K. Tanahashi, and Y. Kawamura** (2001) Work function of GaAs (001) surface obtained by electron counting model. *Journal of Crystal Growth,* **227-228**, 123-126.
149. **Irene Petroff and Viswanathan C. R.** (1971) Calculation of photoelectric emission from tungsten, tantalum, and molybdenum. *Physical Review B,* **4**, 799-815.
150. **Janata, J.** (1991) Chemical modulation of the electron work function. *Analytical Chemistry,* **63**, 2546-50.
151. **Jablonski, A.** and **C. J. Powell** (2002) The electron attenuation length revisited. *Surface Science Reports*, **47**, 33-91.
152. **Janata, J., and Mira Josowicz** (1997) Nernstian and non-Nernstian potentiometry. *Solid State Ionics,* **94**, 209-215.
153. **Jimenez, J., P. Hernandez, and J. A. de Saja (1987)** Optically induced long-lifetime photoconductivity in semi-insulating bulk GaAs. *Physical Review B,* **35**, 3832-3842

154. Johnson, K. B. and W. N. Hansen (1995) An acoustically driven Kelvin probe for work function measurements in gas ambient. *Review of Scientific Instruments*, **66**, 2967-2976.
155. Johnson. E. O (1957) Measurement of minority carrier lifetimes with the surface photovoltage. *Journal of Applied Physics*, 28, 1349-1353.
156. Johnson. E. O **(1958)** Large-signal surface photovoltage studies with Germanium. *Physical Review*, **111**, 153-166.
157. Kamada, M.; Murakami, J.; Tanaka, S.; More, S. D.; Itoh, M.; Fujii, Y. (2000) Photo-induced change of the semiconductor-vacuum interface of p-GaAs(100) studied by photoelectron spectroscopy. *Surface Science*, **454-456**, 525-528.
158. Kamimura T. and M. Stratmann (2001) The influence of chromium on the atmospheric. *Corrosion Science*, 43, 429-447.
159. Kampen, T. U., D. Troost, X. Y. Hou, L. Koenders, and W. Mönch (1991) Surface-photovoltage effects on adsorbate-covered semiconductor surfaces at low temperatures. *Journal of Vacuum Science and Technology B*, **9**, 2095-2099.
160. Kane, E. O. (1962) Theory of photoelectric emission from semiconductors. *Physical Review*, **127**, 131-141.
161. Kaplan T. A (1973) Energy variational principle for a variable number of particles. *Physical Review A*, **7**, 812-814.
162. Kaplan T. A **(2004)** The chemical potential. *Los Alamos National Laboratory Preprint Archive: Condensed matter*, 1-10.
163. Kar S (1975) Determination of Si-metal work function differences by MOS capacitance technique. *Solid-State Electronics,* **18**, 169-181.
164. Katriel J, R. G. Parr, and M. R. Nyden (1981) Concerning the chemical potential of few-electron systems. *Journal of Chemical Physics*, **74**, 2397-2401.
165. Kelvin, L. (1898) Contact Electricity of Metals. *Philosophical Magazine*, **46**, 82.
166. Kindig, N. B. (1967) Effects of band-bending on energy distribution curves in photoemission. *Journal of Applied Physics*, **38**, 3285-3290
167. Knapp A. G (1973) Surface potentials and their measurements by the diode method. *Surface Science*, **34**, 289-316.
168. Knapp J. A and G. J. Lapeyre (1976) Angle-resolved photoemission studies of surface states on (110) GaAs. *Journal of Vacuum Science and Technology*, **13**, 757-760.
169. Koch N, C. Chan, A. Kahn, and J. Schwartz (2003) Lack of thermodynamic equilibrium in conjugated organic molecular thin films. *Physical Review B*, 67, 195330-195335
170. **Koenders L, M, Blömacher, and W. Mönch** (1988) Electronic properties of sulfur adsorbed on cleaved GaAs surfaces. *Journal of Vacuum Science and Technology B*, **6**, 1416-1420.
171. Kremer R. E, M. C. Arikan, J. C. Abele, and J. S. Blakemore (1987) Transient photoconductivity measurement in semi-insulating

GaAs. I. An analog approach. *Journal of Applied Physics*, **62**, 2424-2431.
172. **Krolikowski W. F, and W. E. Spicer** (1970) Photoemission studies of the noble metals, *Physical Review B*, **1**, 478-487.
173. **Kronik, L. and Y. Shapira** (1993) New approach to quantitative surface photovoltage spectroscopy analysis. *Journal of Vacuum Science and Technology A,* **11**, 3081-3084
174. **Kronik L., and Yoram Shapira** (1999) Surface photovoltage phenomena: theory, experiment, and applications. *Surface Science Reports*, **50**, 1-206.
175. **Kronik L., and Yoram Shapira** (2001) Surface photovoltage spectroscopy of semiconductor structuctres : at the cross roads of physics, chemistry and electrical engineering *Surface and Interface Analysi* , **31**, 954 - 967
176. **KuŸminski S, K. Pater, and A. T. Szaynok** (1991) Surface photovoltage investigations of $Cd_{1-x}Mn_xSe$ for 0d"xd"0.05. *Surface Science*, **247**, 90-93.
177. **Lagel, B.; Baikie, I. D.; Petermann, U.** (1999) A novel detection system for defects and chemical contamination in semiconductors based upon the Scanning Kelvin Probe. *Surface Science*, **433-435,** 622-626.
178. **Lagowski Jacek, Andrzej Morawski, and Piotr Edelman** (1992) Non-contact, no wafer preparation deep level transient spectroscopy based on surface photovoltage. *Japanese Journal of Applied Physics, 31,* L1185-L1187; *Lagowski Jacek, Ioan Baltov, and Harry C. Gatos (1973) Surface photovoltage spectroscopy and surface piezoelectric effect in GaAs. Surface Science, 40, 216-226*
179. **Lam Y. W** (1971) Surface-state density and surface potential in MIS capacitors by surface photovoltage measurements. *I. Journal of Physics D: Applied Physics,* **4**, 1370-1375.
180. **Lam Y. W and E. H. Rhoderick** (1971) Surface-state density and surface potential in MIS capacitors by surface photovoltage measurements. II. Journal of Physics D: Applied Physics, **4**, 1376-1389.
181. **Lander, J. J.** and **J. Morrison** (1964) Low-Energy Electron Diffraction Study of Graphite. *Journal of Applied Physics*, **35**, 3593-3598
182. **Lang, N. D** (1981) Interaction between closed-shell systems and metal surfaces. *Physical Review Letters*, **46**, 842-845.
183. **Lang, N.D.** and **W. Kohn** (1971) Theory of metal surfaces: Work function. *Physical. Review* **B 3**, 1215-1223.
184. **Lang, N.D.** and **W. Kohn** (1971) Theory of metal surfaces: Work function. *Physical. Review* **B 3**, 1215-1223.
185. **Larsen P. K, J. H. Neave, and B. A. Joyce** (1981) Angle-resolved photoemission from As-stable GaAs (001) surfaces prepared by MBE. *Journal of Physics C: Solid state Physics*, **14**, 167-192.

186. **Lawrence E. O, and B. Linford** (1930) The effect of intense electric field on the photoelectric properties of metals. *Physical Review*, **36**, 482-497.
187. **Lea, C. and C.H. B. Mee** (1968) Work-function measurements on monolayer films of Uranium on (100), (110), and (113) oriented faces of Tungsten single crystals by photoelectric and contact potential difference techniques. *Journal of Applied Physics,* **39**, 5890-5896.
188. **LeClair, L. R., S. Trajmar, M. A. Khakoo, and J. C. Nickel** (1996) A time-of-flight spectrometer for measuring inelastic to elastic differential cross-section ratios for electron-gas scattering. *Review of Scientific Instruments,* **67**, 1753-1760.
189. **Lee Y. S and W. E. Anderson** (1989) High-barrier height metal-insulator-semiconductor diodes on n-InP. *Journal of Applied Physics*, **65**, 4051-4056.
190. **Leibovitch M, L. Kronik, E. Fefer, L. Burstein, V. Korobov, and Y. Yoram Shapira (1996)** Surface photovoltage of thin films. *Journal of Applied Physics*, **79**, 8549-8556.
191. **Leibovitch. M, L. Kronik, E. Fefer, and Yoram Shapira** (1994) Distinction between surface and bulk states in surface-photovoltage spectroscopy. *Physical Review B*, **50**, 1739-1745.
192. **Liehr M and H. Lüth** (1979) Gas adsorption on cleaved GaAs(110) surfaces studied by surface photovoltage spectroscopy. *Journal of Vacuum Science and Technology*, **16**, 1200-1206.
193. **Lile D. L** (1973) Surface photovoltage and internal photoemission at the anodized InSb surface. Surface Science, 34, 337-367; **Lin Alice L and Richard H. Bube** (1976) Photoelectronic properties of high-resistivity GaAs:Cr. *Journal of Applied Physics*, **47**,1859-1867.
194. **Liu Qiang, Chao Chen, and Harry Ruda** (1993) Surface photovoltage in undopped semi-insulating GaAs. *Journal of Applied Physics*, **74**, 7492-7496.
195. **Liu, Q. and H. E. Ruda** (1997) Role of deep-level trapping on the surface photovoltage of semi-insulating GaAs. *Physical Review B*, **55**, 10541-10548
196. **Logothetis, E. M, H.Holloway, A. J. Varga, and E. Wilkes (1971)** Infrared detection by Schottky barriers in epitaxial PbTe. *Applied Physics Letters,* **19**,318-320.
197. **Ludeke R and L. Esaki** (1975) Electron spectroscopy of GaAs and AlAs surfaces. *Surface Science*, **47**, 132-142.
198. **Lundgren, S. and B. Kasemo** (1995) A high temperature Kelvin probe for flow reactor studies. *Review of Scientific Instruments*, **66**, 3976-3981.
199. **Luo, G. -N.; Yamaguchi, K.; Terai, T.; and Yamawaki, M.** (2001) Influence of space charge on the performance of the Kelvin probe. *Review of Scientific Instruments*, **72**, 2350-2357.
200. **Lüth Hans** (1975) Optical spectroscopy of electronic surface states. *Applied Physics*, **8**, 1-14.

201. **Lyubartsev, A. P, A. A. Martsinovski, S. V. Shevkunov, and P. N. Vorontsov-Velyaminov** (1992) New approach to Monte Carlo calculation of the free energy: Method of expanded ensembles. *Journal of Chemical Physics*, **96**, 1776-1783.
202. **Mackel R., H. Baumgartner, and J. Ren** (1993) The scanning Kelvin microscope. *Review of Scientific Instruments*, **64**, 694-699.
203. **Manaka, T; Ohta, Hideki; Fukuzawa, Masahiro; Iwamoto, Mitsumasa** (2003) Electrostatic properties of polyethylene Langmuir-Blodgett films. *Japanese Journal of Applied Physics, Part 1: Regular Papers, Short Notes & Review Papers*, **42**, 6473-6476.
204. **Many A., Goldstein, Y., and Grover, N. B.,** (1965) Semiconductor Surfaces. North-Holland, Amsterdam.
205. **Mao D, A. Kahn, G. Le Lay, M. Marsi, Y. Hwu, G. Margaritondo, M. Santos, M. Shayegan, L. T. Florez, and J. P. Harbison** (1991) Surface photovoltage and band bending at metal/GaAs interfaces: A contact potential difference and photoemission spectroscopy study. *Journal of Vacuum Science and Technology B*, **9**, 2083-2089.
206. **Mao D, A. Kahn, M. Marsi and G. Margaritondo** (1990) Synchrotron-radiation-induced surface photovoltage on GaAs studied by contact-potential-difference measurements. *Physical Review B*, **42**, 3228-3230.
207. **Martin G. M, J. P. Farges, G. Jacob, J. P. Hallais, and G. Poiblaud** (1980) Compensation mechanisms in GaAs. *Journal of Applied Physics*, **51**, 2840-2852.
208. **Massies J, P. Devoldere, and N. T. Linh** (1979) Work function measurements on MBE GaAs(001) layers. *Journal of Vacuum Science and Technology,* **16**, 1244-1247.
209. **McElheny P. J, J. K. Arch, and S. J. Fonash** (1987) Assessment of the surface-photovoltage diffusion-length measurement. *Applied Physics Letters*, **51**, 1611-1613.
210. **McElheny P. J, J. K. Arch, H. -S. Lin, and S. J. Fonash** (1988) Range of validity of the surface-photovoltage diffusion length measurement: A computer simulation. *Journal of Applied Physics*, **64**, 1254-1270.
211. **McLachlan, N. W., (1951)** Theory of vibrations. Dower Publications, INC., New York.
212. **McInturff D. T, J. M. Woodall, A. C. Warren, N. Braslau, G. D. Pettit, P. D. Kirchner, and M. R. Melloch** (1992) Photoemission spectroscopy of GaAs:As photodiodes. *Applied Physics Letters*, **60**, 448-450.
213. **McMurray, H. N.; Williams, G.** (2002) Probe diameter and probe-specimen distance dependence in the lateral resolution of a scanning Kelvin probe. *Journal of Applied Physics*, **91**, 1673-1679.
214. **Michaelides, A., P. Hu, M. - H. Lee, A. Alavi, and D. A. King** (2003) Resolution of an ancient surface science anomaly: Work function changes induced by N adsorption on W (100). *Physical Review Letters*, **90**, 246103/1-246103/4.

215. **Michaelson H. B.,** (1977) The work function of elements and its peiodicity, *Journal of Applied Physics,* **48**, 4729-4733.
216. **Miller W. R and G. E. Stillman** (1990) Surface Fermi level changes in n-type GaAs determined from Hall-effect measurements. *Applied Physics Letters,* **57**, 2934-2936.
217. **Mönch W** (1993) Semiconductor Surfaces and Interfaces, Springer-Verlag Berlin Heidelberg, Chap 8., 117-136.
218. **Moore A. R and Hong-sheng Lin** (1987) Improvement in the surface photovoltage method of determining diffusion length in thin films of hydrogenated amorphous silicon. *Journal of Applied Physics,* **61**, 4816-4819.
219. **Moss, T. S.** (1955) Photovoltaic and photoconductive theory applied to InSb *Journal of. Electronics Control,* **1**, 126–138.
220. **Mulliken R. S** (1934) A new electroaffinity; Together with data on valence states and on valence ionization potentials and electron affinities. *Journal of Chemical Physics,* **2**, 782-793.
221. **Munakata Chusuke** (1990) An analysis of ac surface photovoltages for obtaining surface recombination velocities in silicon wafers. Semiconductor Science and Technology, **5**, 206-210 ; **Munakata Chusuke and Noriaki Honma** (1987) Saturation of ac surface photovoltages due to photocapacitances in a strongly-inverted oxidized p-type Si wafer. *Japanese Journal of Applied Physics, 26, 564-567; Munakata Chusuke and Shigeru Nishimatsu (1986) Analysis of ac surface photovoltages in a depleted oxidized p-type silicon wafer. Japanese Journal of Applied Physics,* **25**, 807-812.
222. **Murti D. K, L. J. Brillson, and J. H. Slowik** (1982) Photovoltage studies of aluminum-phthalocyanine interfaces. *Journal of Vacuum Science and Technology,* **20**, 233-236.
223. **Nalbach M and H. Kliem** (2000) Contact charging and surface charge measurement using a scanning Kelvin technique. *Physica Status Solidi (a),* **178**, 715-719.
224. **Negoro. N; Anantathanasarn, Sanguan; Hasegawa, Hideki** (2003) Effects of Si deposition on the properties of Ga-rich (4×6) GaAs (001) surfaces. *Journal of Vacuum Science & Technology B,* **21**, 1945-1952.
225. **Noguera, C., D. Spanjaard, D. Jepsen, Y. Ballu, C. Guillot, J. Lecante, J. Paigne, Y. Petroff, R. Pinchaux, P. Thiry, and R. Cinti** (1977) Is the Observed Photoemission Peak near the Fermi Level on the (100) Face of Mo a Surface State?. *Physical Review Letters,* **38**, 1171-1174
226. **Okamoto, Kenji; Sugawara, Yasuhiro; Morita, Seizo** (2003) The imaging mechanism of atomic-scale Kelvin probe force microscopy and its application to atomic-scale force mapping. *Japanese Journal of Applied Physics, Part 1: Regular Papers, Short Notes & Review Papers,* **42**, 7163-7168.

227. **Orlowski B. A** (1988) Electronic surface states investigated by means of photoemission spectroscopy. *Surface Science,* **200**, 144-156.
228. **Palau J. M and J. Bonnet** (1988) Design and performance of a Kelvin probe for the study of topographic work functions. *Journal of Physics E: Scientific instruments,* **21**, 674-679.
229. **Parker James H, Jr., and Roger W. Warren** (1962) Kelvin device to scan large areas for variations in contact potential. *Review of Scientific Instruments,* **33**, 948-950.
230. **Parr R. G, R. A. Donnelly, M Levy, and W. E. Palke** (1978) Electronegativity: The density functional viewpoint. *Journal of Chemical Physics,* **68**, 3801-3807.
231. **Pashley, M. D., K. W. Haberern, W. Friday, J. M. Woodall, and P. D. Kirchner** (1988) Structure of GaAs (001) (2´4)-c(2´8) determined by scanning tunneling microscopy. *Physical Review Letters,* **60**, 2176-2179.
232. **Pashley, M. D., K. W. Haberern, R. M. Feenstra, and P. D. Kirchner** (1993) Different Fermi-level pinning behavior on n- and p-type GaAs(001). *Physical Review B,* **48**, 4612-4615.
233. **Pauling L.** (1932) The nature of the chemical bond. IV. The energy of single bonds and relative electronegativity of atoms. *Journal of American Chemical Society,* **54**, 3570-3582.
234. **Peterson I. R** (1999) Kelvin probe liquid-surface potential sensor. *Review of Scientific Instruments,* **70**, 3418-3424.
235. **Petravic, A., P.N.K. Deenapanray, B. F. Usher, K.J. Kim, and B. Kim** (2003) High-resolution photoemission study of hydrogen interaction with polar and nonpolar GaAs surfaces. *Physical Review B,* **67**, 195325/1-8.
236. **Petroff, I.** and **Viswanathan C. R.** (1971) Calculation of photoelectric emission from tungsten, tantalum, and molybdenum. *Physical Review B,* **4**, 799-815.
237. **Politzer P, R. G. Parr, and D. R. Murphy** (1983) Relationship between atomic chemical potentials, electrostatic potential and covalent radii. *Journal of Chemical Physics,* **79**, 3859-3861.
238. **Qcept Technologies**, Atlanta, USA
239. **Ranke, W.** (1983) Ultraviolet photoelectron spectroscopy investigation of electron affinity and polarity on a cylindrical GaAs single crystal. *Physical Review B,* **27**, 7807-7810.
240. **Reihl, B., J. K. Gimzewski, J. M. Nicholls, and E. Tosatti** (1986) Unoccupied electronic states of graphite as probed by inverse-photoemission and tunneling spectroscopy. *Physical Review B,* **33**, 5770-5773.
241. **Reif F** (1965) Fundamentals of statistical and Thermal Physics, McGraw-Hill Book Company, Singapore.
242. **Renyu, C., K. Miyano, T. Kendelewicz, K. K. Chin, I. Lindau, and W. E. Spicer** (1987) Kinetic study of initial stage band bending at metal

GaAs(110) interfaces. *Journal of Vacuum Science and Technology B*, **5**, 998-1002.
243. **Riviere John C. and Sverre Myhra** (Editors), Handbook of Surface and Interface Analysis, marcel Dekkar, New York 1988.
244. **Rietveld G, N. Y. Chen and D. van der Marel** (1992) Anomalous temperature dependence of the work function in $YBa_2Cu_3O_7$-d. *Physical Review Letters*, **69**, 2578-2581.
245. **Riken Keiki Co. Ltd.**, Azukizawa, Itabashi-ku, Tokyo 174, Japan.
246. **Ritty B, F. Wachtel, R. Manquenouille, F. Ott, and J. B. Donnet** (1982) Conditions necessary to get meaningful measurements from the Kelvin method. *Journal of Physics E: Scientific instruments*, **15**, 310-317.
247. **Roichman Y., and N. Tessler** (2002) Generalized Einstein relation for disordered semiconductors-implications for devise performance. Applied Physics Letters, **80**, 1948-1950.
248. **Rossi F, G. I. Opat, and A. Cimmino** (1992) Modified Kelvin technique for measuring strain-induced contact potentials. *Review of Scientific Instruments*, **63**, 3736-3743.
249. **Rossi F.** (1992) Contact potential measurement: Spacing-dependence errors. *Review of Scientific Instruments*, **63**, 4174-4181.
250. **Rossi F.** (1992) Contact potential measurement: *The preamplifier. Review of Scientific Instruments, 63, 3744-3751.*
251. **Rothschild, A.; Levakov, A.; Shapira, Y.; Ashkenasy, N.; Komem, Y.** (2003) Surface photovoltage spectroscopy study of reduced and oxidized nanocrystalline TiO_2 films. *Surface Science,* **532-535**, 456-460.
252. **Sadewasser, S., Th. Glatzel, M. Rusu, A. Jäger-Waldau, and M. Ch. Lux-Steiner** (2002) High-resolution work function imaging of single grains of semiconductor surfaces. *Applied Physics Letters,* **80**, 2979-2981.
253. **Salmon, L. G., and T. N. Rhodin** (1983) Angle resolved photoemission study of GaAs (100) surfaces grown by molecular beam epitaxy. *Journal of Vacuum Science and Technology B*, **1**, 736-740.
254. **Santos, A. M. D., D. Beliaev, L. M. R. Scolfaro, and J. R. Leite** (1999) Quasi-Fermi level, chemical potential profiles of a semiconductor under illumination. Brazilian Journal of Physics, **29**, 775-778.
255. **Schattke, W.** (1997) Photoemission within and beyond the one-step model. *Progress in Surface Science*, **54**, 211-227.
256. **Schedin, F., R. Warburton, and G. Thornton** (1998) Bolt-on source of spin-polarized electrons for inverse photoemission. *Review of Scientific Instruments,* **69**, 2297-2304
257. **Scheer J. J and J. van Laar** (1963) Photoemission from semiconductor surfaces. *Physics Letters*, **3**, 246-247.

258. Schroder, D. K. (2001) Surface voltage and surface photovoltage: history, theory and applications. *Measurement Science and Technology*, **12**, R16–R31
259. Sebenne C, D. Bolmont, G. Guichar, and M. Balkanski (1975) Surface states from photoemission threshold measurements on a clean, cleaved, Si (111) surface. *Physical Review B*, **12**, 3280-3285.
260. Shafiee A, M. Golshani, and M. G. Mahjani (2002) Bell's theorem and chemical potential. Journal of Physics A, **35**, 8627-8634.
261. Shalish, I., L. Kronik, G. Segal, and Y. Shapira, S. Zamir, B. Meyler, and J. Salzman (2000) Grain-boundary-controlled transport in GaN layers. *Physical Review B,* **61**, 15573-15576
262. Shapira Y, L. J. Brillson, and A. Heller (1984) Origin of surface and metal-induced interface states in InP. *Physical Review B,* **29**, 6824-6832.
263. Shapira Y, L. J. Brillson, and A. Heller Investigation of InP surface and metal interfaces by surface photovoltage and Auger electron spectroscopies. *Journal of Vacuum Science and Technology A*, **1**, 766-770.
264. Sharma, T. K., S. Porwal, R. Kumar, and S. Kumar (2002) Absorption edge determination of thick GaAs wafers using surface photovoltage spectroscopy. *Review of Scientific Instruments,* **73**, 1835-1840.
265. Shegelski M. R. A (1986) New result for the chemical potential of intrinsic semiconductors: Low-temperature break down of Fermi-Dirac distribution function. *Solid state communications*, **58**, 351-354.
266. Sher. A, Y. H. Tsuo, and John A. Moriarty (1980) Si and GaAs photocapacitive MIS infrared detectors. *Journal of Applied Physics*, **51**, 2137-2148.
267. Shikler, R. and Y. Rosenwaks (2000) Near-field surface photovoltage. *Applied Physics Letters,* **77**, 836-838
268. Shimizu, N; Ikeda, Mitsuhisa; Yoshida, Eiji; Murakami, Hideki; Miyazaki, Seiichi; Hirose, Masataka (2000) Charging states of Si quantum dots as detected by AFM/Kelvin probe technique. *Japanese Journal of Applied Physics, Part 1: Regular Papers, Short Notes & Review Papers*, **39**, 2318-2320.
269. Shiota, I., K. Motoya, T. Ohmi, N. Miyamoto, and J. Nishizawa (1977) Auger Characterization of Chemically Etched GaAs Surfaces. *Journal of Electrochemical society,* **124**, 155-157
270. Shockley W., and G. L Pearson (1948) Modulation of conductance of thin films of semi-conductors by surface charges. *Physical Review,* **74**, 232-233.
271. Shockly W., (1939) On the surface states associated with a periodic potential. *Physical Review.* 56, 317-323.
272. Simon Ralph E (1959) Work function of iron surfaces produced by cleavage in vacuum. *Physical Review,* **116**, 613-617.
273. Skriver, H.L., and N. M. Rosengaard (1992) Surface energy and work function of elemental metals. *Physical Review B,* **46**, 7157-7168.

274. **Slater J. C and N. H. Frank** (1947) Electromagnetism, McGraw-Hill, New York.
275. **Smit K, L. Koenders, and W. Mönch** (1989) Adsorption of chlorine and oxygen on cleaved InAs(110) surfaces: *Raman spectroscopy, photoemission, and Kelvin probe measurements. Journal of Vacuum Science and Technology B,* **7**, 888-893.
276. **Smith, N. V.** (1971) Photoelectron Energy Spectra and the Band Structures of the Noble Metals, *Physical Review B,* **3**, 1862- 1878
277. **Smith, N. V., G. K. Wertheim, S. Hirfner, and Morton M. Traum** (1974) Photoemission spectra and band structures of d -band metals. IV. X-ray photoemission spectra and densities of states in Rh, Pd, Ag, Ir, Pt, and Au. *Physical Review B,* **10**, 3197-3206
278. **Smith J. R, F. J. Arlinghaus, and J. G. Gay** (1980) Electronic structure of silver. *Physical Review* B, **22**, 4757-4763; **Smith, J.R.** (1969) Self-consistent theory of electron work functions and surface potential characteristics for selected metals. *Physical. Review,* **181**, 522-529.
279. **Smoluchowski R,** Anisotropy of the Electronic Work Function of Metals, Phys Rev vol 60, pp 661-674 (1941).
280. **Solodky S., Khramtsov A., Baksht T., Leibovitch M., Hava S.,and Shapira Yoram**. (2003) Surface photovoltage spectroscopy of epitaxial structures for high-electron-mobility transistors. *Applied Physics Letters,* **83**(12), 2465-2467.
281. **Spicer W. E, I. Lindau, P. Skeath, C. Y. Su, and Patrick Chye** (1980) Unified mechanism for Schottky-barrier formation and III-V oxide interface states. *Physical Review Letters,* **44**, 420-423; **Spicer W. E, Z. Liliental-Weber, E. Weber, N. Newman, T. Kendelewicz, R. Cao, C. McCants P. Mohowald, K. Miyano, and I. Lindau,** *(1988) The advanced unified defect model for Schottky barrier formation. Journal of Vacuum Science and Technology B,* **6**, 1245-1251.
282. **Spindt C. J, R. S. Besser, R. Cao, K. Miyano, C. R. Helms, and W. E. Spicer** (1989) Photoemission study of the band bending and chemistry of sodium sulfide on GaAs(100). *Applied Physics Letters,* **54**, 1148-1150.
283. **Srivastava G. P** (2000) Theoretical modeling of semiconductor surfaces and interfaces. *Vacuum,* **57**, 121-129.
284. **Statz H and G. A. deMars** (1958) Electrical conduction via slow surface states on semiconductors. *Physical Review,* **111**, 169-182.
285. **Stuckless, J. T, and M. Moskovits** (1989) Enhanced two-photon photoemission from coldly deposited silver films. *Physical Review B,* **40**, 9997-9998.
286. **Sturge M. D** (1962) Optical absorption of gallium arsenide between 0.6 and 2.75 eV. *Physical Review,* 127, 768-773.
287. **Subrahmanyam, A., A. Karuppasamy,** and **C. Suresh Kumar** (2006) Oxygen-sputtered tungsten oxide thin films for enhanced electrochromic properties. *Electrochemical and Solid-State Letters,* **9**, H111-H114

288. **Subrahmanyam, A.; C. Suresh Kumar, K. Muthu Karuppasamy** (2007) A note on fast protonic solid state electrochromic device: NiOx/Ta2O5/WO3-x. *Solar Energy Materials & Solar Cells,* **91**, 62-66
289. **Surplice N. A and R. J. D' Arcy** (1970) A critique of the Kelvin method of measuring work functions. *Journal of Physics E,* **3**, 477-482.
290. **Suresh Kumar, C., A. Subrahmanyam,** and **J. Majhi** (1996) An automated reed-type Kelvin probe for work function and surface photovoltage studies, *Review of Scientific instruments.* **67**, 805
291. **Suzuki, S., C. Bower, Y. Watanabe, and O. Zhou** (2000) Work functions and valence band states of pristine and Cs-intercalated single-walled carbon nanotube bundles. *Applied Physics Letters,* **76**, 4007-4009.
292. **Swank Robert K** (1967) Surface properties of III-V compounds. *Physical Review,* **153**, 844-849.
293. **Szaro L and J. Misiewicz** (1990) The ac surface photovoltage on silicon under sub-band gap illumination. *Physica Status Solidi,* **118**, *185-188.*
294. **Szaro, L., J. Rebisz, and J. Misiewicz** (1999) Surface photovoltage in semiconductors under sub-band-gap illumination: continuous distribution of surface states. *Applied Physics A,* **69**, 409–413
295. **Sze S. M** (1981) Physics of Semiconductor Devices, IInd Edition, *Wiley Eastern Limited, New Delhi.*
296. **Szostak D. J and B. Goldstein** (1984) Photovoltage profiling of hydrogenated amorphous Si solar cells. *Journal of Applied Physics,* **56**, 522-530.
297. **Szuber J** (1988) Electronic properties of the polar GaAs (111) As surface. *Surface Science,* **200**, 157-163.
298. **Szuber J** (2000) New procedure for determination of the interface fermi level position for atomic hydrogen cleaned GaAs(100) surface using photoemission. *Vacuum,* **57**, 209-217.
299. **Taft, E. and L. Apker** (1955) Photoelectric Emission from Polycrystalline Graphite. *Physical Review,* **99**, 1831-1832.
300. **Tamm, Ig.** (1932) A possible kind of electron binding on crystal surface. *Physikalische Szeitschrift der Sowjetunion,* 1, 733-46.
301. **Taylor D. M. and G. F. Bayes** (1994) Calculating the surface potential of unionized monolayers. *Physical Review E,* **49**, 1439-1449.
302. **Tolman R. C** (1938) The Principles of Statistical Mechanics, *Oxford University Press, Great Britain.*
303. **Touskova J., Samochin E., Tousek J., Oswald J., Hulicius E., Pangrac J., Melichar K., and Simecek T.** (2002) Photovoltage spectroscopy of InAs/GaAs quantum dot structures. *Journal of Applied Physics,* **91**, 10103-10106.
304. **Uda M, A. Nakamura, T. Yamamoto and Y. Fujimoto** (1998) Work function of polycrystalline Ag, Au, and Al. *Journal of Electron Spectroscopy and related Phenomena,* **88-91**, 643-648.
305. **Ukah Clement I, Roman V. Kruzelecky, Daria Rcansky, Stefan Zukotynski, and John M. Petz** (1988) In situ work function

measurements in evaporated amorphous silicon. *Journal of Non-Crystalline Solids,* **103**, 131-136.
306. van Laar J and H. Huijser (1976) Contact potential differences for III-V compound surfaces. *Journal of Vacuum Science and Technology,* **13**, 769-772.
307. van Laar J, A. Huijser, and T. L. van Rooy (1977) Electronic surface properties of gallium and indium containing III-V compounds. *Journal of Vacuum Science and Technology,* **14**, 894-898.
308. van Roosbroeck, W. and H. C. Casey, Jr. (1972) Transport in Relaxation Semiconductors. *Physical Review B,* **5**, 2154-2175.
309. **Vancura T., Kicin S., Ihn T., Ensslin K., Bichler M. and Wegscheider W** (2003). Kelvin probe spectroscopy of a two-dimensional electron gas below 300 mK. *Applied Physics Letters,* **83**, 2602-2604.
310. **Vanmaekelbergh, D.** and **L. van Pieterson** (1998) Free Carrier Generation in Semiconductors Induced by Absorption of Subband-Gap Light. *Physical Review Letters,* **80**, 821-824.
311. **Vasquez R. P, B. F. Lewis, and F. J. Grunthaner** (1983) Cleaning chemistry of GaAs(100) and InP(100) substrates for molecular beam epitaxy. *Journal of Vacuum Science and Technology B,* **1**, 791-794.
312. **Vatel. O and Masafumi Tanimoto** (1995) Kelvin probe force microscopy for potential distribution measurement of semiconductor devices. *Journal of Applied Physics*, **77**, 2358-2362.
313. **Vickerman J. C** (Editor), (1997) Surface Analysis-The Principal Technique, John Wiley & Sons Ltd., England.
314. **Vilan A., Abraham Shanzer, and David Cahen (2000)** Molecular control over Au/GaAs diodes, *Nature*, **404**, 166-168.
315. **Villagonzalo C, R. A. Römer and M. Schreiber** (1999) Thermoelectric transport properties in disordered systems near the Anderson transition. *European Physics Journal*. B, **12**, 179-189
316. **Vorburger T V, D. Penn and E.W. Plummer** (1975) Field emission work functions, *Surface Science,* **48** 417-431
317. **Wada M, R. V. Pyle, and J. W. Stearns** (1990) Dependence of H-production upon the work function of a Mo surface in a cesiated hydrogen discharge. *Journal of Applied Physics, 67, 6334-39.*
318. **Weaver J. M. R. and H. K. Wickramasinghe** (1991a) Semiconductor characterization by scanning force microscope surface photovoltage microscopy. *Journal of Vacuum Science and Technology B,* **9**, 1562-1565.
319. **Wieder H. H** (1980) Problems and prospects of compound semiconductor field-effect transistor. *Journal of Vacuum Science and Technology*, **17**, 1009-1018.
320. **Williams Richard** (1962) Surface photovoltage measurements on cadmium sulfide. *Journal of Physics and Chemistry of Solids,* **23**, 1057-1066.

321. **Willis, R. F., B. Feuerbacher, and B. Fitton** (1971) Experimental Investigation of the Band Structure of Graphite. *Physical Review B,* **4**, 2441- 2452.
322. **Wong, K., and V.V. Kresin** (2003) Photoionization threshold shapes of metal clusters. *Journal of Chemical Physics,* **118**, 7141-7143.
323. **Yablonovitch E, C. J. Sandroff, R. Bhat, and T. Gmitter** (1987) Electronic properties of sulfide coated GaAs surfaces. *Applied Physics Letters,* **51**, 439-441.
324. **Yu, P. Y. and M. Cardona** *Fundamentals of Semiconductors,* IIIrd Edition, Springer-Verlag Berlin Heidelberg, Germany, (2001).
325. **Zemansky, M. W, M. M. Abbott, and H. C. Van Ness** (1975) *Basic Engineering Thermodynamics, IInd Edition,* McGraw-Hill Kogakusha, Ltd., Tokyo.
326. **Zhang Xiumiao and Jiatao Song** (1991) The effect of surface recombination on surface photovoltage. *Journal of Applied Physics,* **70**, 4632-4633.
327. **Zisman, W. A.** (1932) A new method of measuring contact potential difference in metals. *Review of Scientific Instruments,* **3**, 367-370.

■■■

Index

A
Adatom-Induced Surface State 9
Adatoms 4, 9
Adsorbate Surfaces 4
Adsorption Kinetics 4
Atomic Chemical Potential 27
Atomic Structure 4

B
Bardeen-Like Barriers 6
Bond Angle 4

C
Canonical Ensemble 25, 28
Chemical Potential 22
Chemisorption 10
Colloidal Graphite 84
Contact Potential Difference (CPD) 32

D
Dangling Bonds 3, 4
Defect States 113
Defects 3
Dember Effect 134
Dember Voltage 116, 136, 137
Density Functional Theory 26, 27
Density Of Surface Defect States 88
Density 103
Diffusion Length Of Minority Carriers 129
Diffusion Length 112
Dipole Layer 6, 11, 34, 101
Dipole Operator 85
Double Layer 33, 42

E
Electric Double Layer 11
Electronegativity 9, 27

F
Fermi Level Pinning 12
Fermi's 'Golden-Rule' 85
Fermi-Level Pinning 6
Field Effect 7
Fringe Field Effects 49
Fringe Fields 71

G
Grand Canonical Ensemble 25

H
Heat Of Formation 6
High Energy Surface 4
Highly Ordered Pyrolytic Graphite (HOPG) 73

I
Inner Work Function 31
Interface Control Layer (ICL) 3
Interface Dipoles 10
Interface States 3, 7
Intrinsic Minimum Surface-State Density 109
Ionicities 9
Ionization Energy 6, 9, 12

K
Kelvin Probe Mean Capacitance 50, 59

M
Minority Carrier Life Time 112
Modulation Index 62, 64, 65
Monolayers 10

N

Negative Electron Affinity (NEA) 3
Non- Parallelism 63

O

Optical Joint Density Of States 85

P

Patch Fields 84
Patches 32, 36, 42, 51, 59
Patchy Surface 49
Patchy 42
Photoionization Capture Cross-
    Section1 117
Polyhedron 42
Positron Work Function 34
Positron 34
Positronium Atom 34

R

Reconstruction 4
Reference Electrode 73
Relaxation 4

S

Schottky Effect 32
Schottky-Like Barrier 6
Spacing Dependence 69
States (Dos) 103
Sticking Coefficient 10
Stray Capacitance 59, 63, 69, 70
Surface Band Bending 12, 19
Surface Band Structure 7
Surface Barrier Heights 5
Surface Dipole 9
Surface Dipole Layer 6, 34, 96, 108
Surface Dipoles. 9, 11, 100
Surface Double Layer 42
Surface Energy 4, 21
Surface Molecule Model 9
Surface Photovoltage Spectra 117
Surface Plasmons 86
Surface Potential 10
Surface Recombination Velocity 136
Surface Reconstruction 7, 10
Surface Space Charge Layer 112
Surface Space Charge 113
Surface State Density 109
Surface States 7, 3, 112,
Surface Stress 23
Surface Tension 23

V

Vibrating Capacitor 46
Virgin Surface 6
Virtual Gap States (Vigs) 10
Virtual Gap States 9
Virtual Ground 59
Volta Effect 32
Volta Potential 39

W

Work Function 21
Wulff Construction 24